Our Sonoran Desert

Bill Broyles

To my wife, Joan Scott,
who shares much blame
for my desert madness.

Contents

"The purpose of the desert is to walk through slowly, and think."
TRAVIS J. EDMONSON, *Thoughts That Didn't Pass*

Desert
More & Less

1\.

LET'S PACK A DAY BAG and take our walking shoes. We'll leave the city behind and try to touch the heart of the most mysterious, beautiful place I know, the Sonoran Desert. You'll need to bring a hat, a water bottle, maybe binoculars, and a camera and notebook, so you can share it all with the folks back home.

When you think of desert, you may picture giant sand dunes. We have them. You may think of treeless plains and dry lakes. We have those. You may think of jagged lavas and steep cinder slopes, fanged serpents and perilous thirst, crimson sunsets and endless vistas. Yes, we have those, too.

Let's take a look. You, too, are welcome to love it, as do we who are blessed to live here. Guidebooks, friends, maps, and your sense of exploration will all enable you to find special desert places of your own.

There's lots of room to choose from. The Sonoran Desert covers 120,000 square miles and runs from 23° to 35° north latitude and from 109° to 116° west longitude—an area as big as the state of Arizona, or nearly as large as Germany or Japan. It's home to about 3,500 species of plants, 500 species of birds, 120 species of mammals, uncounted species of invertebrates. Their challenge? How to live here with more heat and less water.

The name "Sonoran Desert" was popularized by botanist Forrest Shreve, who helped map its boundaries in the first half of the twentieth century. He used the term because it's brief and convenient, it already had some currency among researchers, and much of the desert lies in the Mexican state of Sonora. Some local areas within the Sonoran Desert have earned their own names, such as Colorado Desert, Yuma Desert, Lechuguilla, Vizcaíno, and Gran Desierto de Altar.

But its borders are based on biology, which is forged by heat and cold and rain and geography. This desert differs from others in its combination of plants and animals. Here the saguaro, mesquite, and palo verde are typical plants. For comparison, the Chihuahuan Desert is known for its abundance of yucca, grasslands, and agaves; the Mojave Desert for its Joshua trees; and the Great Basin Desert for its sagebrush.

◀ **The lush Arizona Upland community is rich with cactus, shrubs, and trees. Here we see organ pipe, hedgehog, and cholla cactus in Organ Pipe Cactus National Monument.**
LARRY ULRICH

▶ **Poppies and saguaro skeleton at Picacho Peak, Arizona.**
LARRY LINDAHL

Six habitats

The Sonoran Desert includes six biological communities, which resemble neighborhoods in the same town. When we visit one area, you'll immediately see that it is somewhat different from the others, and yet you'll still know you're in the Sonoran Desert.

The community most readily recognized is the **Arizona Upland,** with its palo verde trees and many cacti such as saguaro, prickly pear, and cholla. It receives the most rainfall—averaging 8 to 16 inches annually—and has the lushest vegetation. Tucson, Arizona, is the most familiar place to see this community. It is called "upland" because most of it stands on slopes and broken plains.

The **Lower Colorado River Valley** is the most expansive community. It ranges from Palm Springs and Needles, California, to Phoenix, Arizona, to Caborca and San Felipe, Mexico, and it includes what some call the Colorado Desert. Typical of the region are large stands of creosote bush and bursage, ironwood trees, and cholla cactus. This also is the driest and hottest of the communities.

The **Central Gulf Coast** is known for its boojums and elephant trees along the Gulf of California. Palmer figs, jatrophas, and palo blanco trees grow among varieties of cacti that include cardóns, organ pipes, and senitas. The Gulf of California islands belong to this community. It includes Loreto and La Paz in Baja California Sur and Guaymas in Sonora. Rain is scant, averaging five inches a year, so trees are seldom leafed.

The **Plains of Sonora** surround Hermosillo, Sonora. The smallest of the six communities, this area is defined by open stands of sun-loving trees, such as velvet mesquite and ironwood. It also has three species of palo verde trees, which are recognizable by their green or blue-green bark; "palo verde" means "green stick" in Spanish. The flame-red flowers of the ocotillo and the metallic blue of the guayacan help to distinguish this "neighborhood" from other desert communities, which are filled with yellow or white blossoms. The Plains have relatively dry winters and wet summers, with 4–12 inches of rain annually. The valleys are broad and the soils deep, and much of the native vegetation has been replaced by irrigated farms and pasture lands.

Two lesser known but equally fascinating communities are the **Magdalena Plain** and the **Vizcaíno** in Baja California, both on the Pacific side of the peninsula. The Magdalena Plain is a drier neighbor of the Vizcaíno. Creeping devil cactus, cardóns, limber bushes, and assorted shrubs dominate, with a few reaching the size of trees.

Although the Vizcaíno is about 9°F cooler in the summer than the other communities, it

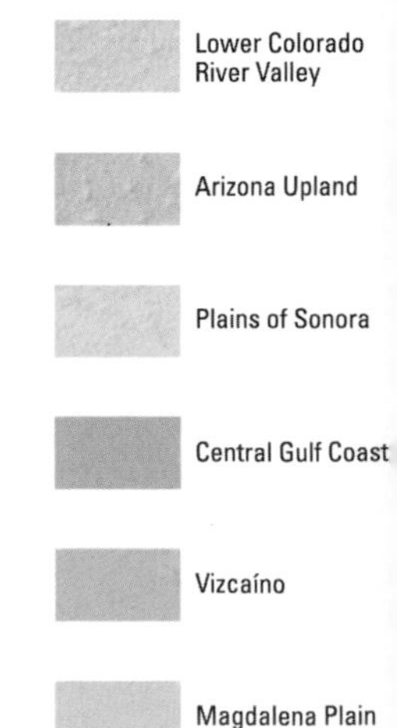

◀ **Coastal dunes, Bahia Gonzaga, Baja California Norte.** MILLS TANDY

▼ **Fog brings moisture to plants in Vizcaíno Desert near Santa Inez, Baja California Norte.** MILLS TANDY

averages only 4–8 inches rain per year. Much of its moisture comes as fog that settles during the night and lasts until midday. As the fog condenses, plants with broad leaves are able to harvest water and funnel it to their roots. Consequently, agaves, yuccas, and dudleyas do very well in the Vizcaíno, and boojums grow especially tall—some up to 75 feet with moss hanging from their branches. This desert region is named for the adjacent bay renowned for whales. The bay was named after Sebastian Vizcaíno, a Spanish sea captain who visited in 1602.

At one time scientists included the thorn forests and shrub thickets around Navajoa and Alamos, Sonora, in a seventh Sonoran Desert community called the Foothills of Sonora, but these areas have been reclassified as Sinaloan thornscrub and dropped from the list of desert habitat.

By small airplane we can visit two locales and see where the edges of all six communities meet. If we fly in a lazy circle around Santa Ana and Benjamin Hill, Sonora, we can see the Arizona Upland, the Plains of Sonora, and the Lower Colorado River Valley habitats. Flying farther to the southwest and across the Gulf of California, we can see the Central Gulf Coast, Vizcaíno, and the Magdalena Plain communities between Santa Rosalia and San Ignacio in Baja California Sur. In a day we can see landscape differences that took decades for scientists to spot by car.

HOW THE DESERT BEGAN

The Greek philosopher Heraclitus said that we can't step into the same river twice. Deserts are the same way. The Sonoran Desert has changed and is changing; the lines and zones are not permanent. Scientific evidence indicates that the Sonoran Desert evolved eight million years ago but that its edges repeatedly have moved north or south depending on the drying and cooling effects of each Ice Age.

During each Ice Age, polar ice caps and continental glaciers locked up enormous quantities of water, causing sea level to fall more than 400 feet below what it is today. Eleven thousand years ago, the northern end of the area now known as Sonoran Desert was cooler and wetter. The plants included juniper, piñon, yucca, and oak. The region resembled oak woodlands and yucca grasslands more closely than anything that we see here today. Relict species of Ice Age plants still can be found in isolated pockets atop desert mountains and in deep canyons.

The modern boundaries of the Sonoran Desert began to take shape about 9,000 years ago, with the current group of plants and animals being in place for the last 4,500 years. How do we know? Pack rats,

isotopes, and old photos, for a start. Researchers such as Paul Martin and Tom Van Devender noticed that pack rats (the genus *Neotoma,* or wood rats) inherit their dens and use them for thousands of generations.

Outside their nests in rock crevices, pack rats heap piles of sticks and cholla balls to fortify their homes against predators such as foxes and snakes. They also urinate at certain spots around their dens, and over thousands of generations the urine glues together layers of pollen, bones, seeds, and leaves that can be identified and dated. Pack-rat middens give us a glimpse 40,000 years into the past.

Scientific examinations of hundreds of nests show that the range of sun-loving saguaros moved north over the last 10,000 years. Creosote bushes, mesquites, ironwoods, and other plants now associated with the Sonoran Desert also spread north as the climate warmed and grew drier. Some plants, including ephedra and bear grass, stayed in place.

Carbon 14 also helps researchers to date plants, bones, and seashells from creatures that lived centuries ago. The process of carbon dating tells us when shells were used as tools and jewelry by the Hohokam. It can date fossils and bones in caves, such as Ventana Cave, where 11,000 years ago woolly mammoths roamed. And it's shown that ironwood trees live as long as 1,200 years and the dead trunk may persist another 1,500 years, because its density and chemistry resists the usual decay agents of termites, beetles, and fungi.

▲ **Desert islands in the Sea of Cortez host dense communities of cardon, organ pipe, senita, and cholla cactus, as seen here on Isla Cholludo.**
George H. H. Huey

◀ **White-throated woodrat eating prickly pear fruit, an important source of both food and moisture.**
C. Allan Morgan

Desert trees are nearly impossible to date by their growth rings, because their wood decays in a few years or because the rings are too close together to distinguish. Cacti have no growth rings. However, evergreen and hardwood trees from mountains near the desert have been dated to 2,000 years before the present. This is a great help in understanding old Indian ruins and campsites.

We also learn from historic accounts such as photos and eyewitness diaries. Some individual plants such as saguaros were photographed a century ago, so photos taken today can be compared to show their growth rates. Some things we'd assume are ephemeral actually may be long-lived. Photos taken in the late nineteenth century show anthills that have prospered since then, and who knows how long before?

These glimpses into the past allow us to see changes in plant and animal communities, draw conclusions about climate change, understand the effects of human habitation, and plan our own future here.

The Enduring Cactus

"If God would choose a plant to represent him, I think he would choose of all plants the cactus… It does not complain when the sun bakes its back or the wind tears it from the cliff or drowns it in dry sand of the desert or when it is thirsty. It protects itself against danger, but it harms no other plant… It is the plant of patience and solitude, love and madness, ugliness and beauty, toughness and gentleness."

—Bryce Courtenay, *The Power of One*

The Sonoran Desert has been called "the succulent desert" in honor of its many fascinating cacti and agaves. Both plants are succulents, which means that they have thick and fleshy water-storing leaves or stems. One difference between the two plants is that most cacti have arms or stems, while agaves have rosettes of sword-shaped leaves.

Cacti evolved over the last 50 million years from close relatives of *Pereskia,* a tropical genus

of plants that grow from southern Mexico to northern Argentina. *Pereskia* look more like trees and shrubs than cacti as we know them, but they have the same distinctive spines and cupped flowers.

Most cacti evolved in the tropics and subtropics and thrive in warm, humid areas. But they are not jungle plants, because they grow slowly and can't compete well with fast-growing plants for light, water, and space. The Sonoran Desert is home to about 300 species of cacti.

Cacti are basically water bags. But don't count on getting a tasty drink from any of them, not even from the fabled barrel cactus said to have furnished water to parched travelers lost in the desert. The slimy fluid inside a cactus is gritty, bitter, and laced with alkaloids and other chemicals meant to discourage man and beast from trying to imbibe. It takes far more effort—and sweat—than it's worth to even moisten your lips, and the cactus does not survive. Without such built-in defenses, cacti would soon be extinct.

Cacti also rely on needle-like spines to protect themselves. Like barbed porcupine quills or fishhooks, they go into flesh easily but are painful to remove. Some are tipped with a chemical irritant that reminds a person or animal to steer clear. Small bristly spines called glochids can be more irritating than the largest spines.

Spines serve several obvious and a few not-so-apparent functions. Above all, they "fence" the cactus from herbivorous animals that would love to eat the succulent cactus pulp. Spines also shade the cactus—ironically, too much sun will burn and possibly kill cacti. Spines reduce wind speeds close to the surface, blunting desiccation and providing a blanket against frost. To shelter the sensitive tips of each stem from the extremes of heat and cold, some species of cactus even grow fuzzy, white caps called trichomes. In one experiment, the tip of a cactus shorn of its spines was 12°F hotter on a summer day and 4°F cooler at night, making it harder for the plant to cope.

Day-blooming cacti—like hedgehogs, prickly pears, and chollas—have glorious flowers in different shades of red and yellow that attract insects. But they have scant fragrance, and they seldom are blue, a color that doesn't appeal to most desert insects. A host of cactus bees, birds, and insects pollinate cacti. Night-blooming cacti, on the other hand, have white flowers—like miniature full moons—and emit seductively stunning fragrances that attract nocturnal pollinators such as white-lined sphinx moths.

The queen-of-the-night cactus grows from large underground tubers. A queen that's three feet tall may have a 40-pound potato-like ball and above ground just a few spindly stalks, which are difficult to spot because they resemble limbs of creosote bushes. It will flower sometime between late May and early July, but each flower blooms only one night and must be fertilized with pollen from a flower on another plant. Its delectable fragrance will leave you with an indelible memory. Botanical gardens host special public evenings when queens-of-the-night are ready to flower. If anyone invites you to see one blooming, drop everything and go.

◀ **Santa Rita prickly pear cactus in bloom. Many animals eat both the fruit and the pads of various prickly pear species.**
RANDY PRENTICE

▶ **Queen of the night cactus flowers open for one night, emitting a memorable fragrance.**
GEORGE H. H. HUEY

Flowers are not the only way cacti reproduce. Chollas and prickly pears have segmented joints that readily break off, much to our pain when we brush against them. Cholla plants are particularly adept at hitching rides on a boot or an animal's leg to start a whole new colony. Where they fall on willing soil, each pad, joint, or even flower pod can sprout adventitious roots and produce a typical plant.

It's hard to remember in our tussles with cacti that some critters actually eat them. Chainfruit cholla fruits are a mainstay in the diet of the Sonoran pronghorn. Pack rats use cholla balls in making their cozy little homes, and they rely on the prickly pear for water. Javelinas eat the prickly pear pads and fruit, spines and all—you have to see it to believe it.

Symbol of the Southwest

The saguaro cactus is a symbol of the Sonoran Desert and, for many people, of the entire Southwest. Its graceful stature and artistic arms make it photogenic and friendly. It appears in movies, logos, and business signs worldwide. A popular bumper sticker bears the salutation, "SAGUARO YOU TODAY?"

Saguaros may reach 60 feet in height—the record is 78 feet—and have as many as 50 arms, though those in the driest soil may have none at all. The arms provide extra space for photosynthesis and flowers, as well as more storage for water.

The pollen and nectar in saguaro flowers is relished by bats, insects, and doves. Each flower may yield 2,000 seeds in a red, pulpy fruit that

▼ An unusual snow graces a saguaro cactus forest, but day-long frosts can kill these desert sentinels.

RANDY PRENTICE

may be the desert's tastiest treat. White-winged doves, rodents, javelinas, and coyotes eat the nutritious fruit and disburse the seeds. The fruit is 70 percent carbohydrate, 30 percent fat, and 10 percent protein.

Papagos [now known as Tohono O'odham] harvesting organ pipe cactus fruit.
LITHOGRAPH BY LT. MICHLER, FROM *Report on the United States and Mexican Boundary Survey, Vol. I*, 1857.

Life is hard for saguaro seeds, produced by the fruit just in time for the summer rainy season. The seeds need summer rain in their first few months in order to sprout, then a series of seasonal rains for two or three years in order to establish themselves. These essential conditions seldom come, so saguaro forests usually have groups of the cacti the same size and age. Because of their tropical heritage, the plants cannot tolerate freezing temperatures for more than a day and do not grow in areas that receive less than two inches of rain in summer. Good crops of saguaros may come only a few times in a century.

A saguaro may live two centuries and be 30 or 40 years old before it's mature enough to bloom. During its life a saguaro may produce 40 million seeds but will be successful if only one grows to replace its parent. Saguaros shun areas where daytime temperatures stay below freezing or where summer rains fall below two inches.

Saguaros are not the largest cacti in the Sonoran Desert, just the most recognizable. Their cousin in the southern Sonoran Desert, the cardón, is actually more massive, but to some minds not as elegant. Other column-shaped cacti include the organ pipe, senita, and creeping devil.

Saguaros and cardóns are more than showpieces—they are sources of food for many birds and animals, they make homes for elf owls and woodpeckers, and they once provided wooden poles for human shelters. They are important nesting and perching sites for raptors such as redtail hawks and ospreys, especially in coastal areas where cacti are taller than the trees.

A saguaro or cardón may contain two tons of water. Ribs or folds give them structural rigidity, provide more surface area for photosynthesis, and allow them to expand and contract depending on how full they are of water. During droughts they can dehydrate up to 80 percent and still survive. Roots may dry, too, and develop a hard cover in order to keep moisture from escaping. Then, when rain does come, the plant will send out new roots—rain roots—within hours and harvest all the water it can. A thirsty saguaro can rehydrate in a few days.

Two other columnar cacti provide food relished by wildlife as well as Native Americans. The organ pipe and senita have many arms, which start from the base. Both produce fruit in June, golf-ball-sized and sweet inside. The organ pipe is known in Spanish as *pitahaya dulce*—sweet pitahaya.

Gila woodpeckers peck nesting holes in saguaros.
TOM VEZO

Staying Safe *in the* Sonoran Desert

"When your canteen reaches half empty, head for home."
RONALD IVES, desert geographer

OLD-TIME DESERT DWELLERS offer advice to visitors and newcomers to make their desert stay more enjoyable. The kinds of things seasoned desert explorers told me, I'll relay to you. We want you to enjoy the desert. It's a very special and fulfilling place, but a few precautions are in order.

How does anyone survive in this desert? The same way people in other climes do. Prepare. Adapt. Think. Before the advent of air conditioning, residents had to cope somehow. Some dodged the heat by building basements, sleeping outside at night, or staying in the shade during the hottest hours of the day. Others watered their dirt floors and hung wet sheets across windows for a primitive form of evaporative cooling. Some sent their families to the seacoast or to cool mountain cabins for the summer.

Follow a few basic rules and you'll be safe and feel at home. Remember, no one is stronger than the summer sun.

- Drink frequently, especially in summer, when you may require two gallons of fluid daily.
- Shield your skin from the sun. Sunburn is not fun, and skin cancer is a real threat. Wear a hat and use sunglasses.
- Never leave children or pets in parked vehicles during the summer. Direct sunlight can within 10 minutes heat the inside of a car to fatal temperatures.
- When hiking even a short distance, take water.
- Use a comb to flick out cholla spines, and use tweezers or rubber cement to remove the tiny barbed spines or glochids from prickly pear. A mat of rubber cement acts as a depilatory and the glochids can be pulled off.
- Don't handle or feed wildlife. They are not pets or toys. Most snakebites happen when people play with poisonous snakes.
- Tread lightly. Desert soil is easily broken and slow to heal.
- If lost, stay put. Make yourself BIG by signaling with a mirror, fire or smoke (but be very careful to control it), making noise, or laying colorful clothes on the ground. Make yourself comfortable by resting in the shade. Stay with the vehicle if you have one. Searchers can spot a car from miles away, but they may not be able to see a person at close range.
- During summer, plan on staying indoors or in the shade from 11 till 3 o'clock. Take a cue from the animals—they stay under cover waiting for the midday heat to break.

As the Greek philosopher Zeno wrote, "When I have provided against heat, thirst, and cold, all else is luxury." We may as well relax and enjoy the luxury of the Sonoran Desert.

Ajo Mountains, Organ Pipe Cactus National Monument.
GEORGE H. H. HUEY

◀ **Native peoples have used agaves for food, soap, textiles, and more. Anza-Borrego Desert State Park.**

CARR CLIFTON

THE ACCOMMODATING AGAVE

Desert plants are like people in an airport—each one has an interesting story to tell if we have time to listen. Do you see that tall stalk? It is a desert agave (*Agave desertii*), and its biography is as complex and typical as a human's. The plant grows one to two feet tall on sandy or gravelly flats in the northern third of the Sonoran Desert. Being a succulent, it stores water and food. It prefers full sun—no shade please, but not the very hottest parts of the desert, since too much sun and heat deplete the stored water. Nor does it like temperatures below 5°F or ground fires. It is tough, but not indestructible.

The desert agave adapts in various ways to its hot environment. It stores water in its leaves and heart. Shallow, wide-spreading roots at an average depth of three inches allow it to harvest light rain, and within a few hours of a rainfall it replenishes itself and begins to grow. Most of its growth occurs within a month of the last quarter-inch of rain. Roots may grow by more than 50 percent during this spurt. They're good at pulling in water, but they also act as gates and don't let water escape from the plant into the ground.

Agaves and cacti are Crassulacean Acid Metabolism—or CAM—plants, which have a special way to make energy from light, water, and nutrients. The CAM process, first noted in members of the *crassulaceae* family of plants, is a way of meeting the challenge of breathing during hot days. The chemical process of photosynthesis requires sunshine, and plants breathe in carbon dioxide and exhale oxygen. But during hot days when a plant opens its stomata to take in carbon dioxide, it loses moisture and faces dehydration. The plant has a dilemma: Grow and risk dehydration, or retain water but not grow. For us, it would be like holding our breath and regulating our metabolism until we can start breathing again.

So a compromise has evolved in CAM plants: They breathe at night, when less moisture is lost, then photosynthesize during the day, with the stomata closed. This process requires more time, so cacti and agaves grow more slowly than many other plants, but at least they don't wilt every hot day. They grow best at 60–66°F.

When the soil stays damp for longer periods of time, the plant switches to a different routine, called C3, and it simultaneously breathes and photosynthesizes during the day. C3 refers to three carbon atoms that are attached during one sequence of the chemical process. This process is more efficient than the CAM mechanism, enabling the plant to grow faster. (A similar but even faster growth process, called C4, is employed by most grasses.) When the soil starts to dry, the agave reverts to its low-speed CAM metabolism. Like all agaves, it blooms only once in its lifetime, and then dies. The price of producing seeds takes the equivalent of a full year's energy. There is no way for the plant to both flower and maintain itself—let alone grow—so it dies after flowering. For all that effort, few if any of that plant's 65,000 seeds reach maturity. Most seeds are eaten by rodents and ants, and the rest lie on hostile ground. Two wet summers in a row are required for a seed to sprout and grow big enough to survive.

Consequently, most new plants stem from nearby plants sending out runners, which become new pups. The "parent" plant may feed water and nutrients to the new plant for 10–15 years until it is fully established and independent. In this strategy, the old plant still lives, and colonies of desert agaves are able to persist. The radial pattern of agave leaves makes the most of sunshine reaching the plant. In summer the stalk, or panicle, shoots up an average of 2.7 inches a day until it reaches 8–20 feet tall. Agaves may wait for decades before the right blooming conditions occur. Flowering depends on the number of wet days in the two preceding years. The yellow flowers have waxy petals and provide copious nectar for the birds, insects, and bats that cross-pollinate the flowers.

Oriole.

FROM *Report on the United States and Mexican Boundary Survey, Vol. II,* 1859. ICTERUS PARISORUM.

HOSPITABLE TREES

The Sonoran Desert also is distinguished from other deserts by its trees, many of them sharing cacti's tropical and subtropical heritage. The most common are legumes, which produce beans. Mesquites, palo verdes, and ironwoods are all important to wildlife for their bean pods and were used by early people for food, fuel, tools, and shelter. Many legumes also "fix" nitrogen in the soil by chemically changing free nitrogen in the air to a compound that plants can use in the ground.

Ironwood trees and saguaros bloom in late spring.

GEORGE H. H. HUEY

This step is crucial for the growth of desert plants because, aside from water, nitrogen is the nutrient in shortest supply. Desert soil may hold less than one percent humus; good garden soil ranges from 25 to 50 percent organic matter.

Trees shelter seedlings of other plants, especially cacti and shrubs. Seedlings depend on the tree's shade and soil enriched by fallen leaves. Saguaro seedlings, for example, can't survive in full sun, even if they receive ample water. The ironwood tree creates a hospitable habitat for at least 160 species of plants. For good reason, such trees are called "nurse trees." But being a nurse plant can be thankless and risky work. Saguaros starting life under a palo verde tree may outlive it and even hasten its death. Saguaro roots lie nearer the soil surface and intercept rainfall before moisture can reach the deeper palo verde roots, so over many seasons the saguaro prospers while the palo verde withers and dies.

Seeds & sprouts

Seeds of most desert plants are cautious and can lie dormant for decades. They respond to a number of factors, which are not clearly understood by researchers. The right timing of rainfall, temperature, amount of rain, and soil moisture are required for germination, and even then, each species of plant responds differently and is difficult to predict.

Most seeds have coats that inhibit sprouting unless washed off by at least a half-inch of rain. This ensures that the ground will be damp enough for them to survive. Some seeds have coats that dissolve in rain but will not leach away if simply placed in water. Others need two or more rains 48 hours apart in order to sprout. Grass seeds wait several days to see if the soil stays damp. The coats of some seeds are eaten away by bacteria, which require prolonged moisture. Saguaro seeds remain viable atop the ground for only a few weeks and require light to germinate. Still other seeds, like those of smoke trees and blue palo verdes, require grinding (such as abrasion in a flash flood) to scuff or nick the outer coating. Other seeds do not sprout in salty soil, so they wait for heavy rains to leach salt out of the soil before they venture into the world. And what an adventure it is!

Surviving the heat

> "Nothing that can survive such shortages should be thought of as ordinary."
> —Molly McKasson, "Molly's Desert Journal"

If we were traveling, we'd have one of three choices for drinking water on our desert trips: We could hoard it in our canteens, gather it from rain and pace ourselves so that it didn't run out, or guzzle what we'd brought with us and hike as fast as we could. Each of these strategies has a parallel in the plant kingdom.

We've already learned that the hoarders of the desert are cacti and agaves, equipped with a natural mechanism for using water efficiently. By contrast, the pacers are trees and shrubs that cope with heat and drought by dropping their leaves in tough times. When they are less able to obtain water, they develop a layer of cells at the petiole that blocks stem water from moving into their leaves—they may even draw water from the leaves back into the stem and store it in the trunk or roots. At that point, the leaves receive no more sap, so they die and fall. These plants grow relatively slowly and invest considerable energy in a woody framework of stems and branches and trunks to support yearly growth and fruit.

▲ **Brittlebush in bloom.**
Carr Clifton

Some pacers, like bursage, not only drop their leaves but they also invest less energy in their flowers and seeds than plants that must attract insects for pollination. Wind and gravity distribute bursage pollen from the stamens to the pistils. The seeds have small burs that cling to fur or cloth to be carried far and wide.

The guzzlers are the seasonal wildflowers that grow ephemerally and have short lives. They live fast, love hard, die young, and leave beautiful memories. They make little attempt to adapt to heat. In the span of a few weeks or months, they sprout from seed, grow, flower, set seed, and die, leaving their legacy of seeds.

The dune sunflower is one of these. Seeds may lie in the sand for decades until the combination of soil moisture, temperature, and nutrition is just right. They grow fast—inches a day—and maximize sunshine by turning their leaves to gather the most light. As in other ephemerals, higher temperatures and stronger sunshine speed chemical reactions and photosynthesis. If the sun overheats them, they can tilt their leaves away.

Desert wildflowers rely on fast growth and vast numbers to outrace the mouths of rabbits, rodents, and even caterpillars. Some of them grow in the spring, and others grow following summer rains. A few, like mallow, are opportunists and grow after any rain. When the soil dries, the plants die back.

The flowers on the datura plant, better known as jimson weed, are huge sweet-scented white funnels that open at night and are pollinated by moths. A plant may grow as big as a bush and may flower all summer long if the soil stays damp. It has big leaves, which may wilt during the day if the roots can't keep it hydrated, but usually it will have renewed its turgor by the next morning.

A few perennial plants combine strategies of hoarding and guzzling. These are the hiders—the vines, gourds, and parasites that have large underground tubers that store water and carbohydrate energy for the next season. In dry seasons the stems and leaves die to the ground. Then when the urge strikes—warmth or rain—the tuber shoots stems up so fast that they are called "runners."

Ajo lilies sprout from bulbs that may lie dormant for several years. In good years, the stem may stand six feet tall and sport several dozen flowers. In years of meager rainfall, the stem may be eight inches tall and valiantly fly only one flower from its stalk. In rainless years, the deep bulb will not show its head above ground. Other plants that grow from bulbs or "corms" include blue-dicks, mariposa, zephyr, and blue sand lilies. More than one desert veteran has been stunned to see fields of lilies where nothing had grown for years—the flowers were awaiting perfect conditions.

Perennial vines include the coyote gourd and climbing milkweed. Coyote gourds produce long vines and large yellow flowers. However, being a tuber has its disadvantages, because some animals can sniff out such plants—even as pigs find truffles. Javelinas are particularly fond of tubers. Climbing milkweed may die back if stressed, but it is persistent and can be seen twined throughout trees and bushes. Although its stems are thin and weak, the stems of several plants may intertwine and support each other like a tripod as they hoist themselves by their own bootstraps to reach limbs up to five feet above the ground. Separately, none of them could reach that far.

Plants and their seeds are building blocks for desert life, supporting herbivorous birds, ants, rodents, and insects, which in turn are preyed on by foxes, hawks, horned lizards, and snakes. And some animals, like coyotes, are omnivorous and take advantage of whatever is on their desert plate.

▼ **Globe mallow and sand verbena flourish before summer's heat.**
Jack Dykinga

Animals
Living Dry

"Happiness is like a butterfly.
The more you chase it, the more it will elude you.
But if you turn your attention to other things,
It comes and softly sits on your shoulder."
— L. Richard Lessor

2.

You ask, "Do any animals live out there?" Let's take a look with our own eyes. We could take a long hike, but instead let's unfold a chair and sit under a big palo verde tree on this *bajada,* the place the mountain slope meets the flat valley. Let the lizards, birds, and inquisitive coyote come to us. Don't worry that you'll miss something—whether we sit or walk, we can't see all of the desert in a lifetime. The fun is in trying.

◀ **Small rodents, such as this Harris' antelope ground squirrel, eat cactus flowers, fruit, and seeds.**
Tom Vezo

▲ **Northern cardinal, a seasonal visitor to the desert.**
Tom Vezo

▶ **A male desert spiny lizard shows off his colors.**
C. Allan Morgan

Lizards & other creatures

Some of our favorite companions will be lizards. Brown, patterned tree lizards hide on the other side of the tree trunk and pretend we don't see them, while slender whiptail lizards spend most of their time stirring the soil at our feet with their pointed snouts in order to find insects. If we stay after dark, we'll see banded geckos. They look frail and nearly transparent, but they survive in a nighttime world of scorpions, centipedes, and spiders. Like some other lizards, geckos use their tails to distract predators. The tail easily breaks off so the lizard can escape—there's some evidence that the lizard can actually drop it off. The tail also stores fat and water.

Large spiny lizards watch us as much as we watch them. They may seem curious, but more likely they're looking to see if we stir up any insects for them to catch. Then they rush down the tree trunk and gobble unwary bugs. Lizards have no teeth, so they swallow their prey whole. Occasionally, they eat flowers. They are quite animated, with head bobs and push-ups done to accentuate their size and vivid coloration, attract mates, and discourage competitors. In winter they'll hibernate in crevices or in burrows with wood rats or tortoises.

Chuckwallas can wedge themselves into crevices to elude predators.
C. Allan Morgan

The biggest lizard in the desert is the chuckwalla, which eats mainly plants and hides in rocky cliffs and piles of boulders. Indians hunted it for food, but the chuckwalla often escaped by wedging itself into crevices and puffing itself up so that pursuers couldn't pull it out. Its flabby skin and dull color also help it to hide and to twist away from predators.

In March the palo verde tree near us will explode with yellow flowers, and a million bees will arrive to feast on the nectar. Many will be natives of the desert—solitary bees, which live in holes and burrows instead of communal hives. These native bees have been called the forgotten pollinators, and we must remember that they have ensured the propagation of seeds since the evolution of, well, palo verde trees. They may range from the world's tiniest bee—the *Perdita minima* that's just one-tenth of an inch long—to large carpenter bees one-and-a-half inches long.

Rock squirrels and antelope ground squirrels climb to the farthest limbs of our palo verde and stuff their cheeks with the green pods and soft beans. Once dry beans hit the ground, cottontail rabbits, javelinas, wood rats and coyotes clean up any leftovers. By then the seeds are rock hard, so animals can digest only the outer pods.

Many animals hide in the shade during midday summer or live underground to avoid the extremes of seasonal or daily temperatures. One study found that when the air temperature in the shade was 108.5° F, the ground surface itself was 160.7° F. But one-and-a-half feet underground, the temperature was a cozy 82.2° F. Some animals hibernate in winter, and others estivate in summer; some do both. Desert tortoises, ground squirrels, and snakes are torpid for much of the year.

A Gila monster awakens from its nap and waddles past our tree. It spends 95 percent of its life underground. How lucky we are to see it today! It is looking for mice, lizards, and especially for quail nests—it loves quail eggs and may eat a dozen at a time, crushing them inside its mouth and swallowing the broken shell and all. Gila monsters are venomous, as are the Mexican beaded lizards living in the southern Sonoran Desert. Both are slow-moving but quick to anger if teased.

The venomous Gila monster lizard is protected by law.
Jim Honcoop

The Gambel's quail, like the saguaro, is a popular emblem of the Sonoran Desert.
Tom Vezo

Female antelope jackrabbits typically give birth to one or two young at a time.
TOM VEZO

Desert cottontail.
C. ALLAN MORGAN

RABBITS

Two kinds of rabbits will hang out near us. Actually, one is a rabbit, and the other is a hare. They may look similar, but each has different strategies for child care. The cottontail rabbit builds a hidden nest in dense brush. The newborn is altricial, meaning it's born blind, furless, and requires its mother's care for several weeks before it can fend for itself. The desert cottontail is the same species found in most of North America. It copes with life

Baby jackrabbits, on the other hand, are precocial. Like other hares, they can see and hop within a few hours of birth. Mom will nurse her litter of one to eight leverets for several weeks, but she may move the nursery regularly so the coyote or bobcat doesn't discover it. Jackrabbits have enormous ears that alert them to predators and serve as radiators by letting excess heat escape. During breeding season, adult males may compete with each other by boxing like kangaroos. Startled jackrabbits escape by jumping in 10- to 20-foot bounds and running in zigzags at up to 45 miles an hour.

BIRDS

Quail chicks are precocial, too. Because quail build their nests on the ground, hidden under bushes and clumps of cacti, they must be ready to move when the chicks are born after 21 days of incubation. The clutch of up to 18 eggs hatches at one time, and the chicks walk behind their parents a few minutes later, their legs looking like pinwheels as they race to keep up with their parents. They learn by imitating the parents' every action, from scratching the ground, to pecking for seeds, to bathing in dust.

Because young quail feed themselves, adult quail can have many more hatchlings than birds that must constantly feed the nestlings. Young quail eat more insects than seeds to get protein for fast growth, but eventually they favor seeds. Adults eat

seeds, seedlings, and nips of leafy plants or flower petals, especially during the spring, because the greens contain vitamin A needed for reproduction. Quail seldom move more than one-quarter mile from where they were hatched.

Both mourning and white-winged doves nest in our palo verde tree, so they can afford to spend more time rearing their young, which are hatched featherless and blind. They may have several broods a year but usually no more than two chicks at one time, since the young must be fed every few hours with predigested pigeon-milk from the

parents' crops. Most doves migrate thousands of miles from southern Mexico to the desert in late spring in order to feast on seeds and saguaro fruit, then return south in the late summer. The cooing of mourning doves and the "who cooks for you?" song of the white-wings fill entire valleys with mellifluous sounds during spring and summer.

We can also find a verdin's nest in our tree. It looks like a small football made of sticks, and the verdin uses it year-round, not just for raising babies. If there is desert mistletoe nearby, phainopeplas also will nest here. The male is iridescent black, and both he and his mate will flutter above the tree to catch insects. Thrashers visit often to search for insects, as does the Gila woodpecker, but neither lives in the palo verde. The thrasher prefers to make a nest in a cholla cactus, and the woodpecker lives in a hollow it makes in a saguaro.

Many desert animals will drink water if it is available, but few require it. Good thing, too, for it is scarce. Gila monsters, Gambel's quail, bull snakes, foxes, and about every other species have been seen to drink surface water. However, they can—and usually do—thrive without it. Coyotes and foxes gain sufficient water from their prey. One study found that coyotes caught more rabbits in summer than winter, indicating they used the catch to allay thirst.

Javelinas

Like many desert animals, javelinas—collared peccaries—are not specially adapted to live in the desert's extreme of heat, cold, and dryness. They are tropical creatures, but they survive quite nicely by changing their behavior and by finding comfortable niches or microclimates. In winter javelinas feed by day and sleep huddled together in shallow caves or even old mine tunnels at night. In summer they switch their routine by feeding at night and early morning to avoid the heat. During the day they nap in the shade of trees and overhanging rocks.

Javelinas are opportunistic feeders and will consume everything from cactus fruits to small animals. They can subsist on cactus for months and derive all the water they need from the prickly pear. They need about one-and-a-half quarts a day, but a cactus meal can provide more than enough water. However, a straight cactus diet doesn't provide enough nutrition over the long haul and causes a buildup of oxalic acid, so javelinas browse other plants, too. They travel in noisy groups of up

◀ **Seldom seen by people, mountain lions have the largest range of any mammal in the New World—from Alaska to Tierra del Fuego.**
E. R. Degginger

▲ **White-winged doves on agave blooms.**
Tom Vezo

▶ **A javelina bristles when startled by danger.**
Mills Tandy

to 20 members. When they come around, we hear them snort and grunt and squeal. Because they have a musk gland, we may smell them before we see them. There is nothing subtle about a javelina.

Hidden life

In late spring, we'll see examples of the bountiful but hidden life of the desert. The patch of drab ground near our palo verde tree may look lifeless, but it holds millions of seeds waiting for the propitious time to germinate. During our stay, seeds will sprout and as many as a thousand plants and grasses may live in one square yard of ground.

Another day we may look out and see a bloom of brightly colored blister beetles feeding on those plants—tens of thousands of red-yellow-and-black beetles eating and mating. They defend themselves with an irritating chemical squeezed out of their pores, so we won't pick one up. Clusters of these beetles denude swaths of plants as they gorge themselves. In a few days, they lay their own eggs in eggs and larvae of grasshoppers or solitary bees. Then the adults die or disappear underground until next year.

And some night we'll look out and see thousands of millipedes grazing the soil crust for mosses, seedlings, and plant debris. They "bloom" after rainy weather, hence their nickname "rain worm." The rest of the year we seldom see any sign of them, for they live underground most of their lives. Despite their name, they don't really have a thousand feet, but it surely seems as if they do.

July and August are my own favorite time, for then the desert is most alive. After it rains, winged ants swarm to mate. Grasshoppers, termites, moths, and butterflies burst into profusion. Lizards, flycatchers, nighthawks, and bats fatten themselves on the bounty.

Tortoises come out of their dens to feed and mate. Surprisingly, they like rocky ground where they can make their dens in shallow rock caves. Tortoises in the eastern parts of the Sonoran Desert are amazingly agile in rocky terrain and on mountain slopes. In the western parts and in the

Some Native American tribes' name for coyotes means "God's dogs."
Jim Honcoop

▲ **Desert tortoises eat a variety of plants and fruits and may live 40 years.**

C. Allan Morgan

Mohave Desert, they prefer sandy flats and dig burrows. By living underground most of the year, they conserve moisture. Their young hatch from eggs and may live 40 years. Like those of saguaros, the ages of adult tortoises in any area tend to cluster around the occasional years of favorable weather conditions—tortoises thrive when the conditions are right for the plants they eat. Janusia vine is their favorite, and even wild tortoises may eat it out of your hand. Because they visually associate the color red with tasty fruits fallen from saguaros, tortoises may travel hundreds of feet to reach any spot of red, be it red fruit, trash paper, or a soda can.

Rattlesnakes

▶ **Sidewinder rattlesnake.**

From *Report on the United States and Mexican Boundary Survey, Vol. II*, 1859. Crotalus cerastus, plate 3.

It's natural to be apprehensive about dangerous critters. You've heard the same stories I have about all the ways to be stung, bitten, hurt, poisoned, maimed, strangled, and wounded, so pretty soon we could be in a needless frenzy of fear. Not to worry. Most human visitors end their stay—and most residents end their year—with no more trauma than the tenderness of sunburn, prick from a cactus spine, or itch from an ant bite.

We haven't seen them yet, but several diamondback rattlesnakes live in the rocky outcrop just up the slope. They emerge several times a year to hunt or mate. We can watch them from a distance where both they and we feel comfortable. When we've crowded them, they usually tell us by buzzing their rattles or heading for cover. Rattlers prefer to avoid large animals, which may kill or eat them. Besides, the snake knows it can't eat something as big as us.

Rattlers hunt mainly at night, because they don't want to be seen by predators such as hawks and much of their prey is nocturnal. Because their metabolism is quite slow, they are very patient and need to eat only three or four times a year. They hunt by following scent trails left by rodents, rabbits, birds, or other snakes. A rattler may coil up and lie in wait for a mouse to return on the scent trail, or it may continue to trail the quarry by inching forward with its body stretched out like a broomstick. Being pit vipers, they rely on their eyes and on heat sensors located in pits on their faces to home in on warm-blooded prey. In a lightning-quick strike, the fangs inject poison from saliva glands. The poison both subdues the prey and begins the process of digestion by breaking down proteins. After all, if you want to eat something your own size, you need to digest it as quickly as possible. One meal of a mouse or rabbit may weigh as much as its predator.

Nearly all rattlesnake bites of humans occur when people try to catch or toy with a snake that wants no part of the amusement. Most victims are young adult males who have drunk too much beer. One study of search-and-rescue personnel who had hiked and worked 40,000 hours in prime rattler terrain recounted no bites. That same jumble of boulders may house tiger rattlesnakes, blacktails, and Mohaves, too. A rattler sheds its skin several times a year and adds a new rattle each time. We may spot young ones in late summer. Born live, they'll be about 8 to 12 inches long and ready to hunt.

In winter they all hibernate in dens deep in the crevices, but on warm days they sun themselves on the front porch.

Other residents of the desert

Occasionally, gray foxes will prowl past our tree. Ghost-like and very hard to see in dim light, they have a natural curiosity and may actually hunt rodents at the end of our flashlight beam. The ubiquitous coyotes may nap under nearby trees like small German shepherds tired after a night's hunt. Their nightly songs are one of our fondest desert memories. They howl and yip to announce finding food, to regroup, and frequently just for the fun of it.

Mule deer will bed near here, but they are so crafty that we seldom see more than their tracks and droppings. At dawn and dusk, they browse among jojoba bushes, mesquites, and fairy-duster. They also graze grasses and seasonal plants such as spurge and filaree. Fawns are born in summer, but they are well camouflaged and precocial. Although mule deer can walk and gallop like other deer, they also have a distinctive bounding gait called "stotting," which looks as though they are riding pogo-sticks. They use stotting to escape predators by changing direction vertically, seeing over shrubs, and maintaining their balance on rough, steep slopes.

White-tailed deer live on the desert's fringes. These are Coues deer, a smaller subspecies. By being smaller, they require less food, which is a good deal if you live in a sparse desert, and their thinner bodies efficiently unload heat. In colder climates, a thicker body helps an animal retain warmth, which helps explain the shape of seals and musk oxen.

We can see a day-full of wildlife from one spot if we are patient and sit quietly. Looking across the valley we see dust devils and rainbows, storms and mirages. We are rewarded by spectacular bolts of lightning splitting the sky and thunder echoing across the land. Our hearts soar with hawks and ravens high above the valley looking for food. From the vantage of one tree we can see a whole living desert.

◀ **Hiking in Aravaipa Canyon, Arizona.**
Tom Brownold

▲ **Desert mule deer, Kofa National Wildlife Refuge.**
Jim Honcoop

Sand

Flats, Dunes, Shores

> "To me the greasewood [creosote bush] is a symbol for health and an example of cheerful existence under adverse circumstances... Were I a poet, I should sing the praise of the modest greasewood of sterling qualities."
>
> — Carl Lumholtz, *New Trails in Mexico*

3.

Between the desert mountain ranges are flat valleys, which may seem to be simpler, plainer, and less interesting than the mysterious canyons and inspiring summits. They at first look like blank spaces between the interesting places. Valleys are the most common feature of the desert, but the word "common" shouldn't lull us into taking them for granted. They are important and can be wondrous. It is the space between notes that makes music and the space between lines that gives us art.

Let's take a walk, not far, but we'll feel as if we've been somewhere. The walking itself is fairly easy, but unless we pick landmarks and navigate with care, we won't be able to find our car on the way back.

◀ **Evening primroses cover sand slopes at Anza-Borrego State Park.**
Carr Clifton

▲ **Roadrunner.**
Tom Vezo

▶ **Horned lizard.**
From *Report on the United States and Mexican Boundary Survey, Vol. II*, 1859 Phrynosoma regale, plate 28.

The rugged creosote bush

Introduce yourself to creosote bush, the most common and widespread of all Sonoran Desert plants. An attractive shrub with waxy, green leaves and yellow flowers, it fills the valley as far as our eyes can see and splashes up the lower slopes of the distant mountains. Sometimes creosote bushes form pure stands, because no other perennial plants can tolerate the dryness and intense sun.

We can't imagine the desert without the creosote bush—it is a keystone species with crucial roles in the ecology and ethnobotany that no other plants fill. It is one of the indicator plants for the Sonoran Desert and it can live a very long time. Comparative photos taken a hundred years ago and today show most of the same plants still living, little changed. If we count clones—one plant spreading its roots to start other stems—a creosote bush may live to be 6,000 years old by the latest estimate.

Creosote bush, like many desert plants, drops some of its leaves when the soil dries out and the temperatures soar. The leaf litter below the plant decomposes to return nutrients to the soil and acts like a mulch to retain soil moisture. This process benefits the many short-lived flowering plants that thrive under creosote bushes in springtime.

However, creosote plants jealously secrete a substance that inhibits other creosote bushes, which helps to explain why they are spaced evenly and widely apart. The drier the soil, the farther these plants are from each other. Creosote bushes can endure more than two years without rain, because they are exceptionally able to pull moisture from dry soil. Part of the species' success lies in its extraordinarily efficient root system that both spreads out close to the surface

to slurp light rains and probes for any moisture deeper in the ground.

Creosote bush is miserly when it comes to spending water. Its resinous leaves themselves are waxy and hold water tightly. They are small so that they don't absorb too much heat and may be dropped in prolonged drought. These are disadvantages in some ways, since growth is very slow, but at least it's steady. The seeds are white, fuzzy, and light enough for wind to scoot them across the ground like blowing sand.

And creosote bush has its own clients which have evolved to blend in quite cryptically. These include the desert clicker grasshopper, green crab spider, and creosote katydid. Because their colors closely match the plant, we must look closely to spot them. At least 22 species of bees feed only on creosote flowers, and 60 species of insects rely on creosote bushes for food or home. There is even a fuzzy, white velvet wasp that looks like a creosote seed, which enhances its ability to dodge predators.

Can you smell the creosote this morning? Whipped by rain from a thunderstorm or kissed by the dew of a winter's morning, that's the desert's most distinctive fragrance. It is clean and pure. It means water and flowers, renewal and tomorrow. It symbolizes persistence, endurance, the affirmation of life—"courage," some would call it.

At certain times of the year, we'll see desert iguanas and round-tailed ground squirrels balancing on the flexible creosote limbs as they harvest flowers and seeds. Most of their time is

▲ **Bursage covers the ground between palo verde trees and saguaros.**

W. Ross Humphreys

◄ **Round-tailed ground squirrel.**

Mills Tandy

spent underground resting and hiding in burrows. The squirrels raise their young in tunnels and rooms they've dug in the sand.

Desert iguanas, like other lizards, lay eggs and do not attend their young. Adults may grow to a foot and a half long and are active in daylight. They feed on creosote leaves and flowers, as well as insects and other lizards. Anyone who has tried to take their photo can attest that they may be the fastest of all lizards, running at 15 miles an hour for a few hundred feet.

We'll avoid that large mound where a colony of kangaroo rats has tunneled around creosote bushes. Other small desert rodents live here in good years, as do snakes, iguanas, and occasionally burrowing owls or badgers. Together they churn up the soil, which allows water and air to reach deeper into the ground, helping the plants grow. But these tunnels and holes make walking difficult and, even if we're careful, our next step may plunge us knee-deep into some rodent's den.

The All-Season Bursage

As we keep walking across the valley, we see a white bursage. It prefers fine, dry soil, and winter rains. In some places creosote bush and bursage make up over 90 percent of all perennial plants. In rockier soil and along runnels that receive summer rains, we find triangle-leaf bursage. Bursages rival creosote bushes in numbers and ability to survive where other plants dry up and blow away. They, too, can drop most of their leaves in severe drought or cold. Their spiny seeds are blown by wind, carried by harvester ants, or caught on animal fur for distribution to new ground. We'll undoubtedly have a few in our socks when we get home.

Bursage is an important nurse plant, sheltering cacti and tree seedlings. In one study of the consequences of bulldozing a patch of desert, bursage took 75 years to re-cover the area. It would take more decades before a crop of barrel, fishhook, or saguaro cactus could grow there. Some of the bursages we're seeing are well over a century old.

Bursage plants have wind-pollinated flowers and bear relatively small seeds, which means they don't need to invest much energy in producing showy petals or big fruits. With sufficient rainfall, they can bloom in any season and several times annually. It takes a rainfall of at least an inch to induce the seeds to sprout. For bursages, as for many desert plants, microrrhyzal fungi in the ground are needed to produce healthy soil and plants.

Bobcat with a ground squirrel.
E. R. Degginger

Boojum trees and cactus near Cataviña, Baja California Norte.
Mills Tandy

Occasionally we'll cross cinder flats, remnants of explosive volcanoes. They form rich soil, and in Mexico stands of organ pipe or senita cactus grow on them. Closer to the rivers, we'll find gravel benches of former channels. In some places, the ground is strewn with chalcedony, chert, and agate stones, polished from their trip downriver.

Paved with surprises

As we walk through the desert, we notice a surface that looks as though a mason laid a single layer of stones on the ground to make a floor. This surface is called desert pavement and results from several processes. When wind blows soil away, the heavier pebbles and rocks remain on top. Animals, too, can bring rocks to the surface when they dig burrows. Another explanation is that repeated wetting and drying of clays in the soil "floats" the stones to the surface. Weathering of rocks by alternating heating and cooling helps whittle rocks down to pebbles. Regardless of the process, pavements are relatively stable and long-lived.

Pavements are renowned for surprises. They look like furnace slag and in summer are as hot as a stovetop. But because so few plants can grow there, they make ideal places for ephemeral plants to sprout in cool, wet winters and then flower in the spring. With little competition to bother them, desert sunflowers, chicory, Ajo lilies, four-spots, and dozens of other species flourish. For the rest of the year—and maybe for decades—seeds lie in the cracks between the rocks awaiting the right rains. We won't soon forget the sight of acres of yellow and white flowers against the black pavement.

Some plants live only a short while but leave skeletons to remind us of their beauty, or at least their presence. Buckwheats, also known as desert trumpet and skeleton weed, leave knee-high stalks to twist and blow in the wind. Primroses leave bleached skeletons of "birdcages" or "devil's lanterns" that persist for years.

One preposterous plant of the pavements looks dead most of the year. It is spiny herb, actually a member of the buckwheat family. In spring it is vivid green and has spine-tipped leaves and yellow flowers the size of pinheads. A spiny herb two inches tall may have a root ten inches long. After seeding, the plants dry, leaving spiny carcasses to remind us of glory gone and to serve as nurse plants for the next generation of seedlings.

Older pavements and basalt boulders will have a black or brown sheen of desert varnish, which is the result of microscopic bacteria collecting dust. The dust contains manganese and iron, which give the varnish its metallic shine. It forms only in extremely hostile environments, for it can neither compete with lichen nor survive in temperate rain.

After a while, we come to a central wash lined by trees. The wash has steep banks cut by swift, erosive runoff. The banks are soft, with few rocks and little gravel. Because of low rainfall, there isn't enough water to carry boulders, rocks, or even gravel very far from the mountains, so most soil in the central valley is clay, silt, and sand.

Wherever rainwater drains and forms runnels, ironwood, mesquite, and palo verde trees may grow, providing perches and nesting space for kestrels and owls, which hunt in the flats. Thrashers and cactus wrens nest in nearby chollas, living off the abundant insects. In the driest lands, trees grow only along these little ditches.

After storms that bring several inches of rain, the barest-looking ground may burst with amaranth, mallow, and summer poppies as far as the eye can see. Thickets grow chest-high and obliterate trails and roads.

The blue palo verde is royalty among desert trees. Growing to a height of 40 feet, it produces yellow explosions of flowers in spring. Only the riparian cottonwood trees grow larger in this desert. Visitors soon learn to distinguish blue palo verde from mesquite, ironwood, and foothills palo verde simply by color. Mesquite is dark green, ironwood a silver green; blue palo verde is blue-green, and foothills palo verde is lime green.

Another big tree along the washes—big by desert standards—is desert willow. It is related to the catalpa, not the willow, but it really is a bignonia. Desert willow produces beautiful purplish, whitish flowers in spring.

◀ **The Kofa Mountains rise abruptly from the plains below.**
Jim Stimson

▶ **Bark scorpion carrying her babies.**
C. Allan Morgan

It loses its leaves during winter and may drop them again in summer until rains come. Its roots may extend 50 feet down and, with supplemental water, a tree's branches may grow three feet per year.

Over there we see a cluster of crucifixion thorn trees (*Castella emoryi*), named after Lt. William H. Emory, who helped survey the U.S.–Mexico border in 1854. They look like exceptionally stout palo verdes, but we'll seldom see many of them, even in one place. Though they grow well in dry places, they are not widely scattered. The problem is that their heavy seeds may hang on the tree for five to seven years. Then they just plop down. The seeds have no wings to blow in the wind; they have no hooks to catch on passersby; they don't grow close enough to stream channels to catch rides on flash floods, and few creatures eat them. It's no wonder they grow close to the parent tree. They are so unusual that one patch of them west of El Centro, California, is specially protected.

On the other hand, the heavy palo verde seeds, like mesquite, acacia, and ironwood seeds, are widely spread because they grow in tasty bean pods that many animals love to eat. The rock-hard, undigested beans are carried in the animal's alimentary canal and then deposited—and fertilized—in scat as the animal makes its daily rounds.

A Living "Apartment House"

Perhaps the most startling feature of the valleys is the least obvious and least photogenic. It is the "bare" ground between the bushes and trees. Instead of a soil crust formed by hardened clay, cemented particles of soil, or bare dirt, we find a crust that is alive. It's the skin of the desert, made from a diverse living community of microscopic fungi, algae, cyanobacteria, lichens, and animals, such as protozoa, nematodes, rotifers, and microarthropods. It is a biologic apartment house. Mites, isopods, snails, mole crickets, tardigrades, termites, millipedes, and ants all feed on the residents.

This soil crust is an inch or less thick, so it is very fragile. Tire tracks and footprints may remain visible for a century. The crust traps moisture in the soil and—because its filament "roots" bind together the soil particles, even grains of sand—the crust inhibits erosion. On windy days, dust may roil off an unpaved road while the natural land near the road is dust-free. This is crucial to other plants, which could be left naked when the soil is removed, or be buried when the dust returns to earth. The crust also provides nutrients, especially nitrogen, and acts as a seed bed.

Sometimes we'll find the crust growing under a chunk of quartz. Like a little greenhouse, the clear rock lets light through and keeps moisture in the soil. If the crust is brown, dry, and dormant, we can dampen it with water from our canteen and within half an hour, it usually will turn green, a sign its photosynthesis has begun. In a land of sporadic and short rainfall, it must be able to awaken within hours and use the water for growth and reproduction. The crust grows very slowly, and a bared patch may take a century to heal.

After a recent shower or even heavy dew, we need to look for crimson-colored velvet mites. They are about the size of a pencil eraser. They live below ground in sandy soils and prey on ants and termites, but occasionally they "bloom" on the surface looking for mates or food in such numbers that it looks like spontaneous generation. For this reason they are sometimes called "sand babies."

Ghosts of the Desert

Pronghorns may live in these creosote flats. They look like antelope, but they really constitute their own genus and aren't closely related to African antelopes. They rely on keen eyesight and fleet feet to detect and elude coyotes, mountain lions, bobcats, and humans. Adult pronghorns can sprint at 60 miles an hour. Although fawns are born ready to walk and can run short distances within a few hours, they rely on knee-high grass and shrubs to conceal themselves their first few months.

▶ **Senita cactus with sand verbena, daisies, and poppies in the Vizcaíno Desert.**
George H. H. Huey

◀ **Desert centipede.**
Tom Vezo

In summer pronghorns may move out of the flats into the foothills in search of cholla fruit and summer plants. Both bucks and does have horns, but the males have a black band around the neck and larger horns. Pronghorn herds are sparse and difficult to locate. For this reason, they are known as "ghosts of the desert." The Sonoran Desert is home to two subspecies, the Sonoran pronghorn, which lives in southwestern Arizona and northwestern Sonora, and the peninsular pronghorn that lives in the Vizcaíno plains of Baja California.

Shifting sands

"I succumbed to the desert as soon as I saw it..."
—Antoine de Saint-Exupéry,
Wind, Sand and Stars

Deserts mean sand, and in places the wind piles it into huge, moving dunes. Walking through dunes is like sailing on the crests and troughs of a storm-tossed, quartz sea. There is no straight line, only curves, sensuously smooth curves. The sand is soft, almost silky. The dunes shimmer with life of their own.

As you'd expect, plants trying to grow on moving dunes face some special problems. Blowing sand may bury a plant or expose its roots. Nothing grows on the steepest slopes, but plants may grab a footing on the fringes where sand moves more slowly. Giant dune buckwheat has extra long roots as insurance against being uncovered. Its root crown may be perched several feet in the air, or the trunk may lie on the ground tethered to a root pivoting in the sand 20 feet away.

There is even a subspecies of dune creosote bush that grows tall and erect instead of bushy, helping it stay above drifting sand. If it is unearthed and topples over, it's quick to send out new stems and several new bushes may sprout from a fallen one.

The parasitic sandfood, a tuber that may grow to 50 pounds, feeds on other plants, so it needs no leaves of its own. All that peeks above the ground is a small disk-shaped flower in the springtime. Native Americans used the tuber for moisture and food.

◀ **Beetle tracks on drifting sands, Cabeza Prieta National Wildlife Refuge.**
Jack Dykinga

▼ **Ajo lilies and sand verbenas thrive in Mohawk Dunes, Yuma County.**
Jack Dykinga

Blowing sand is especially hard on plant seedlings. Besides facing abrasive sand capable of sandblasting telephone poles into toothpicks, seedlings may be covered so they can't get light or excavated so they can't reach nutrients or moisture. Dunes efficiently hold moisture several feet down, but until a seedling's root is long enough to reach that far, it is at the mercy of the fast-drying surface.

Some plants, such as dune buckwheat, giant Spanish needles, Wiggins' croton, Peirson's locoweed, and sandfood, grow only on moving dunes. But, inevitably, plants will tame the moving dunes in a succession from adventurous pioneer plants to sedate creosote bush, bursage, and galleta grass.

We sometimes find hummocks of spindly mesquites and ephedra in the dunes. Since their roots stabilize the sand, hummocks are havens for animals such as foxes, rodents, lizards, and snakes. The sand around them is blank paper for animal tracks. The small footprints of a black Pinacate beetle, the padded feet and tail sweeps of kangaroo rats, the looping S's of sidewinder rattlesnakes all tell stories of a night's activities. Sometimes we can track an animal back to its den, or see where it met its demise in the talons of an owl. The next wind will wipe the slate clean, opening new chapters in the dune story.

Kangaroo rats

Kangaroo rats are esteemed residents of stable dunes. They have long, tufted tails and hop like kangaroos. The tail gives them balance and acts as a rudder during their pell-mell rushes to escape predators. They gather seeds in fur-lined cheek pouches and store them underground in extensive tunnels that they dig with their hind feet. They may have storage rooms, nurseries, and sleeping chambers, as well as half a dozen escape routes. They do not truly hibernate, but they rely on stores of seeds to feed them through the winter.

Because they metabolize seeds and make their own water, kangaroo rats never need to drink, which is a good thing in a land of so little rain. They have an enlarged, inflated mastoid bone (tympanic bullae) surrounding the inner ear bones, which allows them to hear the low-frequency sounds of snakes crawling on sand and the gliding wings of owls.

Anyone who has driven near Palm Springs, California, knows about dunes. On windy days the sand blasts across the roadway, pitting windshields and scouring paint. Winds are particularly intense here when Santa Ana winds gust through passes in the mountains. The venturi effect speeds up the wind and easily moves the unsettled sand. Keeping roads and railroad tracks clear is a full-time job, despite the drift fences that have been erected.

The **Mohawk Dunes** east of Yuma, Arizona, are low hills which in rainy years may be covered with a solid carpet of purple sand verbena and white ajo lilies. Fringe-toed lizards, horned lizards, foxes, and an occasional Sonoran pronghorn may be seen there. The Cactus Plain dunes near Bouse, Arizona, are stable, making them an especially good place to observe dune life. Like the Algodones and Mohawk dunes, their supply of sand has been cut off from the Colorado River.

The **Yuma Dunes** south of Yuma are mostly closed to the public but comprise an important homeland for the rare flat-tailed horned lizards, which eat ants. These aren't just any ants. These are large harvester ants whose vise-like jaws inflict painful bites. They also are armed with a stinger that delivers a startling and itchy jab; even horned toads suffer if they aren't careful.

The **Gran Desierto de Altar** of northwestern Sonora is the largest sand sea or "erg" in North America. The highest of the dunes reach 200 feet tall, and the troughs between the dunes are big enough to cup a football stadium. They run from Yuma nearly to Puerto Peñasco, Sonora. At times cold winds from the north, known as El Norte winds, blow for days on end, driving sand like the lash of a whip. Even the intrepid Padre Eusebio Kino, the 17th-century Jesuit missionary who was among the first Europeans to visit the area, detoured far out of his way to avoid these "continuous, violent, and most pestiferous winds" that moved the sand and shaped the dunes.

But few times in life are as tranquil as sitting atop a dune at sunset, watching day become night and admiring the infinite stars.

Along the Gulf of California

"For whatever we lose (like a you or a me)
it's always ourselves we find in the sea"
— e e cummings,
"maggie and millie and molly and may"

Beyond the dunes, we recognize the land as desert. It is brown, dry, rocky, a bit forbidding, without water, without shade. But if we look closely, we can see signs of an ocean. That nest in the enormous cardón cactus belongs to an osprey, a hawk that feeds on fish. The mysterious boojum trees, which look like upside-down carrots, sport moss watered by marine fogs. Limber bushes have a bonsai look with windswept branches.

▲ **California sea lions, Gulf of California.**
C. Allan Morgan

We're standing on the eroded terrace of a former sea level, and along the shore are broken blocks of sea shells cemented together long ago. The smell of the air is laden with salt, and the sounds of raucous sea gulls fill our ears. Low waves roll in, churning the sand. We see a needlefish jump out of the water, and a shoal of pelicans flying single file dip their bills to the surface in hopes of catching sardines and mullet. A sea lion barks from a solitary reef.

Unlike other North American deserts, the Sonoran Desert is not landlocked. This one has its own gulf. About six million years ago cataclysmic faults split the peninsula of Baja California away from mainland Mexico, opening a rift now known as the Gulf of California. Those faults—the San Andreas, Cerro Prieto, and others—still carry southern California and Baja California northwest from the mainland at the rate of two inches a year.

▶ **The upper Gulf of California, also known as the Sea of Cortez, is famous for its extreme tides.**
Jim Stimson

The Gulf of California is rich in sea life. Whales, porpoises, giant manta rays, sea bass, turtles, sharks, shrimp, and hundreds of species of fish make this one of the world's most diverse and rewarding places to snorkel and scuba dive. And, because of the rich sea life, birds are bountiful, and frigates, boobies, pelicans, terns, and gulls can be seen from shore.

Ancient desert people learned to use this ocean. The Seri Indians took refuge on Tiburón, largest of the Gulf's islands. They lived on land but navigated the Gulf in preposterously small boats made from bunches of carrizo reeds lashed together. With these they traveled to other islands, moving families, trading with other groups of Seri, and harvesting native plants and wildlife.

Traditionally they ate eel grass, fish, and turtles from the sea. They adeptly speared triggerfish and manta rays with pointed lances. They harpooned sea turtles and netted sierra and corvina. They made skirts out of pelican feathers and skinned sea lions for hides to make blankets. Ashore they hunted deer, bighorn, rabbits, and anything that was edible or useful. Now they make prized carvings from ironwood and drive pickup trucks.

The presence of an ocean does little to slake the thirst of the land. Weather over the Gulf is desert-like, with rain being sporadic and sparse. A few of the islands, such as San Pedro Mártir, were noted for their harvestable guano deposits used to make fertilizer and gun powder. Unwashed by rain, these centuries-old collections of bird droppings remained where they fell in the arid climate. The islands of San Esteban, San Lorenzo, and Ángel de la Guarda (Guardian Angel Island) have no permanent water. Tiburón has a few springs and waterholes, but no streams.

Occasional storms, known as *chubascos* and *tormentas,* hit the coast, but they are sporadic and hit so hard that much of the rainfall is lost to runoff. Hurricanes sometimes move up the Gulf from the eastern Pacific and unleash catastrophic flooding. Up to six inches of rain may fall in a few hours, and the resulting flash floods roar through canyons and villages. Much of the desert topography is shaped by these sporadic cataclysmic events that cut arroyo banks, move boulders, uproot trees, change channels, and erode the land.

▼ **Common dolphin, Gulf of California.**
C. Allan Morgan

▶ **Sea-kayakers, near Isla Espíritu Santo, Baja California, enjoy both the desert and the sea.**
Greg Vaughn

Because the Gulf is narrow with only one outlet, it has exceptionally high and low tides at its northern end during the new and old moons. In the Colorado River delta, the spring tide may vary 30 feet twice a day.

These extreme fluctuations lead to an impressive phenomenon known as the tidal bore—or *el burro.* When the incoming tide runs faster and higher than the river, it creates a wave of water up to seven feet tall moving upstream at four to eight miles an hour. The wave breaks constantly as it rolls upriver, roaring as loud as a freight train. Frenzies of gulls and terns feed as it churns fish to the surface. It can sink or ground boats. After the bore races several miles upstream, it runs out of energy and once again the river wins. Other great bores can be found on the Amazon, Seine, and Firth of Forth.

Before major dams were placed on the Colorado River to control flooding, generate electricity, and provide agricultural irrigation water, the river annually flooded into the upper Gulf, bringing key nutrients and fresh water. In the shallow water of the delta lives the world's smallest porpoise. Called the *vaquita,* it is formally known as the Gulf of California harbor porpoise. Adults grow to about four or five feet and weigh less than 120 pounds. They number only a few hundred, and these are seldom seen in the turbid water. Some drown when they become entangled in fishing nets. Nowadays the river seldom reaches the ocean, and sea life is increasingly pressed to cope. Current research stresses the urgency to somehow replenish those old flows.

The Gulf of California offers sunny beaches and a fascinating variety of creatures and plants. We never tire of walking the strand looking for glass floats from fishing nets, driftwood, whale bones, and sea shells. Because the Gulf is relatively placid, kayaking, snorkeling, and fishing are popular activities, convincing us that the worst day at the beach beats the best day at the office. In this desert by the sea, we can watch bighorn sheep in the morning and blue whales in the afternoon.

Rocks From Inner Earth

"We could see a big hole of such depth that it caused terror and fear."

Juan Mateo Manje,
Luz de Tierra Incógnita

4.

The ground rises gently. Palo verde trees and saguaros stand mute and the ground crunches underfoot. A zebra-tailed lizard speeds across our path and a cactus wren searches for food. On the horizon dark, rounded peaks look years distant. Suddenly, the ground opens in front of us. Another step would plunge us into a pit nearly 800 feet deep and one mile across. We stand frozen with astonishment, mouths agape. Nothing prepared us for this. It is a volcano called Crater Elegante, where violent explosions starting 150,000 years ago blew out ash, tuff, and basalt bombs the size of basketballs.

This is the largest of the enormous craters in the Pinacate lava field of northwestern Sonora. Crater Elegante is a maar, or a circular crater that occurs where volcanic vents superheat ground water and cause explosive steam. Here the Sonoyta River fed the groundwater, and explosions and lava flows then knocked the river 40 miles off its course. Where once the river emptied into Adair Bay, it now drains into an estuary east of Puerto Peñasco, Sonora, on those rare occasions when it flows at all.

The dark peaks towering beyond the crater are the eroded walls of Volcán Santa Clara, a shield volcano reminiscent of the Hawaiian Islands. The summit, Pinacate Peak, was named for the darkling eleodes beetle that stands on its head when bothered and squirts a noxious liquid. Flows of black and red basalt look baked by the sun, but they were cooked into submission by a much greater fire inside the earth and came spilling from vents on the mountain. Magma, molten rock at 2,100°F, flowed down its sides, making nearly impassable reefs 10 to 30 feet thick. Now dormant, this shield volcano was active for about the last million years. Padre Eusebio Kino named it Volcán Santa Clara.

Some of its flows are pahoehoe, flat bands of relatively smooth basalt that resemble semicircles

◀ **Basalt overlaying granite tells a geologic story. Here, a many-headed barrel cactus grows beneath Tordillo Butte in the Cabeza Prieta Mountains.**
Jack Dykinga

▲ **Elegante Crater, a maar volcano in Pinacate Biosphere Reserve, Sonora, Mexico.**
W. Ross Humphreys

▶ **Zebra-tailed lizard.**
Jim Honcoop

of hardened chocolate. Other flows are known as *aa* (pronounced "ah ah"), loose basalt slag jagged enough to shred leather boots. In some places we find lava tubes where the onrushing lava cooled on the outside and the core kept flowing and left a hollow rock tube big enough to walk in.

The Spanish had a word for such places: *malpaís,* which translates to "very bad country."

◀ **Brittlebush blooms in the midst of a basalt lava flow, Pinacate Biosphere Reserve, Sonora, Mexico.**

George H. H. Huey

Padre Eusebio Kino climbed to the summit in 1699 and observed that Baja California was connected to the mainland—it was not an island as some explorers had thought. Maps had to be redrawn. From the summit, we too can see the Gulf and Baja. Hundreds of cinder cones dot the landscape. In spring their ashen slopes are a riot of flowers, with yellow brittlebush, red ocotillos, and blue lupine.

Because the Sonoran Desert lies along the Pacific Rim of active vulcanism, it has other dormant volcanoes and flows, such as the Tres Vírgenes volcanic region of Baja California, the Sentinel Plain of Arizona, and the Volcanic Hills of southeastern California.

In geologic terms, much of the Sonoran Desert geology is caused by the Pacific Plate slipping northward along the edge of the North American Plate, dragging Baja California and southwestern California with it for about 180 miles. The boundary of these plates is seismically very active, and major fault lines such as the San Andreas and Cerro Prieto faults are well known to residents of the lower Colorado River.

The subterranean forces that split the Baja peninsula from the mainland are enormous and powerful beyond our imagination. Most of this work has been done in the last 15 million years. Today the earthquakes have subsided and are unusual events in most parts of the Sonoran Desert.

Where the San Andreas fault comes ashore below Mexicali, Mexico, the ground bubbles and spurts mud from water heated deep inside the earth. Other mudpots burble near the Salton Sea. Near Mexicali thermal wells tap the underground brine to drive steam-turbine generators for electricity. These wells at Cerro Prieto use water at 750° to 1,100°F to produce 400 megawatts of electricity yearly.

Mountains in the Desert

Desert mountains do not bother with foothills; they rise from level plain to steep slope with little transition. Most desert mountains are granite, basalt, gneiss, rhyolite, or schist. Born of fire, and heat, and pressure, they are young, rugged, and jagged, with knife-edge crests. The rough edges are not yet worn off; in fact, they are still being cut. They owe their genesis to vulcanism and faulting. This is what geologists call basin-and-range terrain, where plates of the earth's crust collide or rip apart. The earth is stretched and pulled, and then punctured by volcanic eruptions.

Two great episodes of mountain building have occurred in the Sonoran Desert. The first lasted from 200 million years ago to 15 million years back. During that time the Pacific Plate butted against the North American Plate and slid under it. Mountains were pushed up and volcanoes dotted the landscape.

Fifteen million years ago the Pacific Plate changed direction, moving northwestward, and began to pull the crust apart, creating enormous faults and rifts, splitting the mountains apart and dropping whole valleys. A few volcanoes remained active where the Earth's mantle was thin and

▶ **Granite boulders, Anza-Borrego Desert State Park.**

Scott T. Smith

fractured, such as along the San Andreas fault. The major faults trend northwest, so most of the mountain ranges also trend northwest.

The traumatic faulting also gives rise to some extreme escarpments. Mount San Jacinto near Indio, California, rises more than 10,000 feet above the Salton Basin, and in Baja California the massive Sierra San Pedro Mártir towers 10,000 feet above the Gulf of California. Tall cliffs of volcanic rock are landmarks of the Superstition Mountains, Kofa Mountains, and Ajo Mountains.

And some mountains have strange stories to tell. Perhaps the most mind boggling are the Tucson Mountains, west of that city. They once were part of an enormous crater, 15 miles in diameter on top of the neighboring and taller Santa Catalina mountain range. Sometime between 17 and 30 million years ago the land uplifted, and the caldera slid 20 miles downhill to its present location.

> "Proof
> of your existence? There is nothing
> but."
>
> — Franz Wright, "Year One"

From a distance all mountains look forbidding, impossible, aloof. What look like blank slopes turn out to be steep canyons and knife-edge ridges leading to precipitous cliffs. The summit is guarded by loose rock, thorny thickets, and our own frailty.

But it is there. We are here. And the only cure for curiosity is to see for ourselves what mystery awaits in the summit. We'll lace tight our shoes. And pack extra water. And bring our patience and persistence, for after all, what is a marathon but one step at a time? There probably is no foot trail up the mountain, unless we count the paths of coyotes and deer, who may not be headed where we want to go.

◀ **Desert bighorn ram and ewe.**

C. Allan Morgan

Canyon dwellers

Among the most surprising residents of some desert canyons are two species of palm trees. They require running water or springs, further indication of the Sonoran Desert's tropical heritage. The California fan palm (*Washingtonia filifera*) reaches 45 feet tall and may live two centuries. Places such as Palm Springs, Palm Desert, and Twentynine Palms, California, and Palm Canyon in the Kofa Mountains of Arizona, are named for these. In the mountains southwest of Mexicali, Baja California, and near Guaymas, Sonora, we find Mexican fan palms (*Washingtonia robusta*). Several species of blue palm occur in Baja California.

The shaggy thatch of fronds skirting the palm trunk may occasionally burn, but evidently fire seldom hurts mature palms because their water-carrying vesicles are spread throughout the trunk, not just below the bark. Also, fire suppresses other

▼ **Native California fan palm trees, such as these in Anza-Borrego Desert State Park, reflect the Sonoran Desert's tropical heritage.**

Carr Clifton

trees that might compete with the palms. The date fruits are prized by birds and coyotes, which in turn distribute the seeds widely. Optimal conditions for seedlings need to occur only every century for populations to remain viable.

High on the ridge we spot a herd of desert bighorn sheep. They live in mountains throughout the Sonoran Desert. Adult rams have massive horns, which they use to defend themselves from predators and from other rams in head-butting contests over mating rights. Ewes, which have short, thin horns, travel in bands with lambs and other ewes much of the year.

Baboquivari Peak is a "sky island" that rises above the Sonoran Desert.
WALT ANDERSON

Notice how some species of plants grow on warmer, south-facing slopes while others seem to prefer cooler, north-facing slopes or shadows behind boulders or ridges. Saguaros, elephant trees, and ocotillos need the extra warmth and light while yuccas, grasses, and agaves like less heat. We find pools of water amid lush stands of catclaw, golden-eye, trixis, and tree-beargrass. Upstream we spot a spring trickling out of a crack in the rock and flowing briefly downstream.

"HOW TALL IS THE DESERT?"

The summits of higher desert ranges act as sky islands, supporting plants not found lower down. For example, the summit ridges of the Kofa, Sand Tank, and Ajo Mountains host Chihuahua juniper, hopbush, rosewood, banana yucca, barberry, and scrub oak. Because air temperatures decline 3°F for each 1,000-foot gain in elevation, the tops are cooler than the valley floors. They also intercept more clouds, so they receive slightly more rainfall. Heavy canyon shade shelters plants that otherwise might desiccate in the full sun of the valley floor.

Actually, the childlike question "How tall is the desert?" is a good one. Above about 4,250 feet, we find few desert plants. In a few mountain ranges, such as the Santa Catalinas, Santa Rosas, Bradshaws, and at Kitt Peak, we can drive to the top. Along the way we'll pass through the Sonoran Desert and into other life zones, such as oak woodlands and pine forests, just as we would if we drove from the desert floor to Canada. At Mount San Jacinto, we can take the tramway to the summit and marvel at the vista and plants.

We will be tired when we finally step onto the summit, but we will have earned the right to sit a while and enjoy the view. We can see range after range. A redtail hawk circles below. Why climb clear up here? Now we remember. We're alive.

"Climate is what we expect; weather is what we get."
— Sign on weather-watcher's desk

Sun
The Edge of Thirst

"But it's a dry heat."
— An old saying of desert dwellers

5.

◀ Sonoran Desert National Monument boasts one of the finest stands of saguaro cacti anywhere.
Jack Dykinga

▲ In parts of the desert that receive less rainfall, the saguaros have fewer arms (Castle Dune Mountains, Kofa National Wildlife Refuge and Wilderness).
Laurence Parent

▼ Teddy bear chollas, organ pipe cacti, and saguaros, Organ Pipe Cactus National Monument.
Larry Ulrich

Step outside in June and you'll agree that this is a hot desert. Summer temperatures range up to 125° F, with some cities having as many as 100 consecutive days over 100° F. To some this is too much of a good thing, but we learn to cope by drinking extra fluids, dressing to protect ourselves from sunburn, and staying indoors or at least in the shade during midday. Other hot deserts include the Sahara and Kalahari, but on some sweltering summer days a visit to a cold desert—the Gobi or even Antarctica—sounds very inviting.

Deserts and semi-deserts cover 31 percent of Earth's land surface. When most of us think of desert, we imagine very little rain and sometimes no rain. Precipitation in the Sonoran Desert is erratic and unpredictable, ranging from zero in some places to 15 inches a year in others. Averages run 3 to 12 inches, enough to fill a cup or a tall glass of water.

But deserts really are defined by evaporation exceeding rainfall. In the Sonoran Desert this may exceed 10 feet a year. This means that our glass of water will dry up long before we've slaked our thirst. Dropping at the rate of a half-inch a day in June, lakes, rivers, swimming pools, or our glass of water deplete dramatically if not replenished. This dryness severely limits life most days of the year.

Some experts categorize desert zones into hyper-deserts, deserts, and semi-arid deserts, depending upon the amount and regularity of precipitation. The hyper-arid deserts receive less than one inch of precipitation annually in no regular season, while deserts receive less than 10 inches per year and may have seasons. Semi-arid deserts get under two feet of precipitation. The Sonoran Desert has places in all three categories, but yearly averages mean little when all the rain may come at one time, or one year's rain may vary fivefold from the next.

Five seasons of the desert

Although old-timers joke that there are two seasons here—hot and hotter—the Sonoran Desert really has five: spring, hot foresummer, rainy summer, fall, and rainy winter. Two wet seasons are real blessings...if they come. They provide two growing seasons for plants, fill natural water holes twice a year, and make for very interesting weather.

Spring is dry but has warm days. It lasts from February to early May. Afternoons may become windy, and dust devils, a signature of the desert, are common. These whirlwinds are generated when the sun rapidly heats the ground, and the fast-rising air starts turning like a small tornado or cyclone. Some afternoons we can see half a dozen dust devils swirling across a valley. They may last up to an hour and tower to several thousand feet above the ground.

Most are smaller, but others leave dust on our eyes, rip roofs off houses, or veer cars off the highway. Dust devils can turn either clockwise or counterclockwise. One desert sport is to track them and determine their direction of spin.

Hot foresummer starts in May and runs until the rainy season begins in June or July. It is very hot—up to 125°F—and very dry, sometimes with a relative humidity as low as five percent. High pressure domes often hang over Nevada and block moisture from entering the Southwest region.

Hot foresummer days cause air to rise rapidly off the ground and rocks. For days on end, the hot air rises, cools, and wrings out any moisture. Clouds form and may drop what looks like rain, but it is actually virga and evaporates before reaching the ground.

Rainy summer may announce its arrival with a thunderclap. One day, after a series of false alarms, the clouds and heat and humidity reach the trigger point (technically, the dew point reaches 54°F for three successive days), and things finally happen. Lightning! Thunder! A raindrop or two test the dust. Then the sky opens with the best music the desert can hear. A deluge can drop as much as six inches in one episode.

Since hard desert soils don't absorb rain rapidly, much of the rain may run off, causing ferocious flash floods that turn dry arroyos into raging torrents in minutes. Cars that try to cross arroyos may be swept downstream, and bridges may wash out. Water even a foot deep can tumble a car downstream or sweep away a wading person. Signs reading "DIP" or "DO NOT ENTER WHEN FLOODED" should be heeded. Generally, the rushing water subsides in a few hours. Patience is the best policy; desert residents soon learn to wait until the water subsides before trying to cross.

◀ **Dawn in the Cataviña region, Baja California Norte, Mexico.**
Jim Stimson

▶ **Fishhook pincushion cactus.**
Jack Dykinga

This wet monsoon summer occurs mainly in July and August. The rains are caused by warm, humid fronts coming off the Gulf of Mexico or the Gulf of California. Occasionally, the downdraft from a thunderstorm results in a microburst with winds exceeding 125 miles an hour. Wide swaths of trees, ocotillos, and saguaros may be leveled as if pushed over by a giant's hand. On rare occasions, thunderstorms can spawn tornadoes.

Another phenomenon of desert rains is sheet flooding. We may find flat areas where the ground looks as if it had been vacuumed. Every twig and pebble has been hosed clean by a torrent of water several inches deep spread over the entire area. The rodent population may be low for a few years, since many animals drowned and others perished from hunger after their underground stores of food got wet and rotted. Later, adventurous newcomers and prolific survivors will recolonize the area.

Fall months, particularly September and October, seem like a chance to rest up from the frenzy of summer. Temperatures are mild and days grow shorter. Ants and rodents store food for the impending winter. Hiking and camping are pleasant.

Winter in the Sonoran Desert can be cool, but temperatures seldom fall below freezing for more than a few hours. Snow is so rare that local residents can recite the years it snowed at their house and without hesitation will show family photos of themselves building snowmen and throwing snowballs.

▶ **Summer flash flood, Organ Pipe Cactus National Monument.**
George H. H. Huey

Most cacti, and especially the saguaro, can tolerate overnight freezes but must warm up during the day or they suffer frostbite. For this reason, saguaros at the northern edges of the Sonoran Desert grow on south-facing slopes. Occasionally winter storms roll in from Alaska and that pestiferous wind—El Norte—may blow for days on end. A traditional Tohono O'odham prayer asks, "Please stop the wind."

Winter rains come as frontal storms. Banks of clouds roll in from massive low-pressure areas formed in the northern Pacific Ocean, and cool-weather storms may last for days. Their rains generally have a more lasting effect, since they are gentler, with less runoff, and evaporation is less in the cool months.

Seasons and rainfall patterns in the Sonoran Desert are driven mainly by the North American high-pressure zone, which is similar to meteorological zones in other deserts around the world. When equatorial air is heated by the sun and rises, it circulates toward the poles. On the way, it cools and dries in the upper atmosphere, and then warms as it falls earthward between 20 and 35 degrees latitude. This is why most of the planet's deserts occur in the horse latitudes along the Tropic of Cancer and Capricorn.

This high pressure usually excludes low-pressure fronts and currents that bring moisture. But sometimes the jet stream shifts or strong fronts of moisture move down from Alaska or in from the Pacific and push the high-pressure dome off its center. These westerly winds circulate worldwide in

a band from 30 to 60 degrees latitude. In summer, moist air over the Gulf of Mexico blows over the Sierra Madre Mountains in Mexico or up the Gulf of California. In winter, frontal storms blow down from the Pacific Northwest.

Sun & rain

Both summer and winter rains in the Sonoran Desert are affected by water temperatures in the Pacific Ocean as far away as Peru and Australia. The Southern Oscillation, a changing of ocean temperatures, causes variations in rainfall patterns known as El Niño and La Niña.

In the Sonoran Desert, El Niño years are characterized by more rain in winter and less in summer. El Niño may bring floods, such as the famous flood of 1983, which washed out highway bridges, eroded river banks, disrupted traffic, spoiled farmland, and flooded archaeological sites that had withstood rains over the last thousand years. La Niña years are drier in winter and wetter in late spring and summer. Small differences in the balance between summer and winter rains can cause major differences in the lives of plants, animals, and people.

Generally, if soil is dry, a storm must drop more than seven-tenths of an inch of rain in an area for the water to run off. Few storms drop this amount, so flash floods or sheet-wash runoff do not occur regularly.

The balance of Sonoran Desert rain—roughly 50 percent in summer and 50 percent in winter—helps make it the lushest and biologically richest desert in the world. The Mojave Desert receives most of its precipitation in the winter, so its plants and animals must adapt to long, dry summers. Winter rainfall also dominates in the Great Basin Desert, but it often comes as snow and is accompanied by subfreezing temperatures, so that the warm growing season is relatively dry. In contrast, the Chihuahuan Desert receives its rain mainly in the summer.

◀ **A precious desert water hole, Saguaro National Park (East).**
JACK DYKINGA

Each of these four deserts is caused by a rain shadow, the drop in precipitation that occurs on the leeward side of a mountain range. Moist air coming eastward off the Pacific Ocean is wrung dry over the Sierra Nevada of central California by the time it reaches the Mojave and Sonoran Deserts, and the Sierra Madre Mountains keep moisture from the Gulf of Mexico from reaching the Chihuahuan Desert. The Sonoran is hemmed in by the Sierra Madre, Sierra Nevada, Sierra Giganta, and the Colorado Plateau, all of which reduce the amount of rainfall it gets each year. Parts of the Sonoran Desert along the Pacific Ocean receive fog but seldom rain.

If the desert seems bright, it is. Because of lower humidity and clearer air, it receives more sunshine and, because it has less vegetation, there is more sunlight reflected off the ground. Sand and light-colored rocks can be dazzling, even in winter.

Light also plays tricks on us. Mirages create visions of water where we know there is none. Differential heating of air by the sun causes light to shimmer, which we see as "water." The hotter the ground, the bigger the "lake." This effect also causes early-morning mirages, where distant mountains look like castles or appear to rise up in places they really aren't. In some cases, refracted sunlight behind a mountain range can make it loom over the horizon, so we can "see" it indirectly.

Many of us prefer desert weather and have moved to the Southwest from harsher climates where we grew weary of shoveling snow, mowing lawns, or enduring months on end without seeing the sun. Even for a visit, desert weather is refreshing.

Natural water buckets

Rain in the Sonoran Desert is a memorable event. Here it comes during two seasons a year, winter and summer—or at least it is supposed to. Rain is highly variable, and one locale may receive several deluges while a place nearby receives not a drop. Much rainfall in the desert evaporates, some is used by plants, some percolates into the ground, and if there is any left, it runs off.

In the bedrock of the mountains, runoff scours out basins, which catch and hold water. These are known by the Spanish word *tinaja,* meaning "earthen jar or bucket." They range from bowls holding water a few days to enormous scour holes 20 feet deep, 30 feet across, and big enough to hold water through several years even if no more rain comes.

As you'd expect, the locations of these tinajas were well-known not only to wildlife, but to prehistoric Native Americans and pioneers. Generally they are in canyon bottoms and can be located by following old trails which converge from the valley. Predators may go there as much to hunt other animals as to drink. Since the water isn't purified, we won't sample its taste. Besides, some dove or deer may need it.

These tinajas may be dry for years, but when they fill with water, miraculous things happen. Within a few days an assortment of water life sprouts as if by magic. Algae grow and bloom. Tiny eggs of fairy shrimp, horseshoe shrimp, water beetles, and mosquitoes hatch. Dragonflies and damselflies will arrive to lay eggs, but then they will wander dozens of miles from water to feed on tiny flying insects such as gnats.

The names of tinajas and springs have been handed down from generation to generation for millennia, since they literally are life's blood for people. Well-known tinajas in Anza-Borrego State Park in California include Sheep Tanks, Smoke Tree Canyon, and Waterfall Canyon. White Tanks near Phoenix are in a regional park. Arizona's Kofa Mountains have a number of tinajas. Along the infamous El Camino del Diablo—the Devil's Highway—Tinaja del Tule and Tinajas Altas provided water to thirsty argonauts traveling in the California gold rush in 1849.

In valleys with clay soil, the sudden arrival of rain may cause water to pocket for a while in a *charco,* a natural clay-bottomed "pan." A charco provides a home for amphibians—especially spadefoot toads—that seem to appear magically from nowhere to mate, lay eggs, hatch, and spurt from polliwogs to hopping toads within 10 days. They race to grow before the pool dries up. Tadpoles usually eat algae, but to accelerate the growth of at least a few individuals and ensure that they reach maturity, some tadpoles may turn carnivorous and eat their siblings. This gives them extra protein for fast growth.

Adult spadefoots spend most of their lives burrowed several feet underground in a mucous cocoon, which allows them to absorb oxygen from the soil but prevents desiccation. They respond to the sound of summer thunderstorms that signal the advent of rain, a chance to mate, and a feast of insects. Their mating calls, which sound like bleating sheep, can be heard a mile away. A toad may devour enough bugs in one night to last it through the next year. Like spadefoots, red-spotted toads and giant Colorado River toads may also seem to appear magically out of thin air.

Toads also like dry lakes known as *playas*, which means "shores" or "beaches" in Spanish. Dry lakes may have been real lakes long ago in wetter climates, but today they are shallow, dry dust bowls waiting for infrequent rains. Because these playas lack outlets, water may stand a few inches deep over hundreds of acres after heavy rains. Few plants grow on playas, because over millennia minerals precipitated out of the water to form an inhospitable hardpan, hence their other common name: alkali flat.

Much like Saharan oases, the renowned water holes of the desert are springs. Towns such as Palm

◀ *El Camino del Diablo, Arizona, 1985*
JAY DUSARD

▲ **Cattails grow along the Bill Williams River, Havasu National Wildlife Refuge.**
GEORGE H. H. HUEY

Springs are named for canyons where palm trees thrive on spring water trickling from faults caused by the San Andreas earthquake zone. A famous spring in Organ Pipe Cactus National Monument is Quitobaquito, home of desert pupfish and Sonoran mud turtles.

RIVERS

"NO FISHING FROM BRIDGE"
— sign on a bridge over the Hassayampa River

The sign on the bridge suggests that we're crossing a river on our journey through the Sonoran Desert, but our feet are dry and gravel crunches underfoot. The riverbed is dry as a bone. One local legend says that people who drink from the Hassayampa are forever unable to tell the truth. Newcomers who hear anyone call this a river may agree.

Like most desert rivers, washes, or arroyos, the Hassayampa is dry most of the year, and we can walk from bank to bank. But following heavy rains, it becomes a raging torrent, and we would need to swim or paddle for our lives. In 1890 the river burst a reservoir dam and swept scores of miners to their deaths.

There are no desert rivers, at least not any born in the desert. What we think of as real rivers begin high in mountains many miles upstream, where snowfall melts in spring and provides a steady supply of water. The mighty **Colorado River** begins 1,700 miles upstream in the Wind River Mountains of Wyoming, where glaciers and heavy snow linger year-round. It is known as the American Nile, because it too passes through a desert and has a fertile delta.

The **Gila River** begins in the distant mountains of New Mexico. Even before we built dams on the Gila, evaporation took its toll, and the Gila sometimes dried up before meeting the Colorado near Yuma, Arizona. Today it seldom flows all the way. In Sonora the **Río Yaqui** and **Río Fuerte** start in the Sierra Madre but only occasionally reach the sea. Much water is lost to irrigation, evaporation, and human consumption.

In other places, rivers may disappear underground and flow inside layers of gravel and sand. Later, if pushed to the surface by underground walls of rock or impervious soil, water may reappear as pools or short streams. These pools may hold stranded schools of fish and provide excellent hunting for herons.

Like modern residents of the Southwest, the early Hohokam farmed by diverting rivers into canals and ditches to raise melons, beans, and maize. They lived in the Salt River and Santa Cruz River Valleys of southern Arizona, but their civilization foundered when massive floods in the 1380s radically downcut their main canals, leaving the side ditches and precious fields high and dry. The floods were followed by decades of drought and by soil too salty to grow crops. The Hohokam abandoned irrigation farming and went to small-scale plots where seeds were grown in charcos and where brush dams spread water on the soil next to arroyos. *Hohokam* means "people who have gone." Their descendants likely are the modern Akimel, Ak Chin, Tohono O'odham, and Hia-ced O'odham living near Phoenix and Tucson.

The Colorado has major dams designed to prevent flooding, produce hydroelectric power, and provide water for irrigation and drinking. Boulder, Parker, Imperial, and Laguna Dams control and redistribute the once-wild river which seasonally flooded as high as 300,000 cubic feet per second and inundated wide swaths of floodplain. Dams on the Salt River impound Roosevelt, Apache, and Canyon Lakes, and San Carlos Dam tames the Gila River. Vast agricultural areas such as the El Centro, Salt River, Yuma, and Coachella Valleys rely on these reservoirs, which also provide fishing, water skiing, and boating.

Before the dams, the Colorado delta had gallery forests of immense cottonwoods and mesquites. The marshes and flats of the delta were rich in wildlife. Aldo Leopold called it "the green lagoon," a place so wild and fertile with deer, birds, and animals that jaguar lived there. After visiting the delta in 1894, Edgar Alexander Mearns wrote: "The tide creeks and broad bays about our camp were swarming with waterfowl, which were nowhere else seen in so great abundance. Pelicans, cormorants, geese, ducks, cranes, herons, and small waders almost covered the shores and bays; the sky was lined with their ever-changing geometrical figures, and the air resounded with their winnowing wing-strokes and clanging voices, not only during the day, but through most of the night."

The lower Colorado River still has rich pockets of wildlife. Several wildlife refuges—Imperial, Havasu, Cibola, Cienega del Indio, and Cienega Santa Clara—are important bird sanctuaries for shore and migratory birds. The river has fish, too. Striped bass, bass, and catfish are supplanting native fish, and many species have declined in

◀ **The Gila River, lined by mesquite, cottonwood, and saltcedar trees.**
BERNADETTE HEATH

▶ **Raccoons are frequent residents along Sonoran Desert rivers and streams.**
JIM HONCOOP

▶ **Western grebe on the Colorado River.**
RANDY PRENTICE

numbers, including the Colorado River squawfish, a minnow growing to six feet long and 80 pounds. The river flow is barely enough to supply the people, cities, farms, factories and, yes, wildlife that depend on it.

Lakes

Natural lakes in the Sonoran Desert are few. The most famous is the **Salton Sea**, a basin 228 feet below sea level in southeastern California. It is a remnant of Lake Cahuilla, a geologically ancient trough along the San Andreas fault. At one time it was an arm of the Gulf of California, but sediments from the Colorado River formed a silt bar and dammed the inlet. Waves cut benches that can be seen hundreds of feet above today's water level. Southwest of Mexicali, another arm of the trough forms Laguna Salada, which occasionally fills with seawater.

Eventually, Lake Cahuilla evaporated, but several times during wetter climates it was refilled by rivers coming out of now-arid Death Valley and by the Colorado River. From at least five hundred years ago until 1905, it was a dry salt pan called the Salton Sink. But an irrigation ditch near Mexicali broke in 1905, and for two years the Colorado River poured into the trough. The new lake was christened the Salton Sea. At its southern end are several salton domes, remnants of sea salt that evaporated in shallow bays eons ago, and mudpots, where steam bubbles soupy mud out of the ground, like some playful toy geyser.

◀ **Cormorants at sunset, Salton Sea.**
Tom Vezo

▼ **Dams have created desert reservoirs such as Roosevelt Lake, pictured here.**
Larry Ulrich

Allegorical Storm Representing Evening with a View from Nature Painted on the Gila River, Arizona, **circa 1855.**
HENRY CHEEVER PRATT (1803–1880)

The Salton Sea is now about 35 miles long and 15 miles wide, holding seven million acre-feet. Its deepest point is 51 feet with an average depth of 30 feet. Because the lake lacked fish, tilapia, corvina, croaker, mullet, and striped bass have been introduced. Salt and freshwater marshes along the shore provide habitat for numerous migratory birds that stop or winter at the Salton Sea. It is a good place to observe Canada geese, snow geese, American avocets, plovers, and green-winged teals. Resident birds include Yuma clapper rails, brown pelicans, and peregrine falcons.

Farm-water runoff and occasional rains help maintain the current water level, but the annual evaporation of more than a million acre-feet leaves salts behind. With no outlet and only intermittent inflow, the Salton Sea is growing increasingly saline, much like the Dead Sea between Israel and Jordan.

Dams on the Colorado, Salt, Gila, and Bill Williams Rivers create water habitats for wildlife and are playgrounds for humans in the desert. When desert sailors tire of swimming, water skiing, fishing, or boating, they can hike side-canyons resplendent with jojoba and saguaros and watch desert bighorn sheep drinking at water's edge. The Apache Trail to lakes on the Salt River is renowned for its springtime flower show of lupines, poppies, and owl clover. These lakes also provide drinking water, irrigation, and hydroelectric power to the region's cities.

DAMS, AQUIFERS & CANALS

Some modern desert cities rely on groundwater from subterranean aquifers, but there's a price: the aquifer is never replenished, since there is so little rain in this arid land. Wells must be deepened periodically in order to reach the falling water table, and eventually the ground subsides, sometimes forming giant cracks. In some places in central Arizona, square miles of ground have covertly subsided 30 feet over the last century, leaving enormous fissures and cracks.

Increasingly cities such as Phoenix, Las Vegas, and Los Angeles are dependent upon water imported by canals. Ultimately, water will limit how many of us can live in the Sonoran Desert and will determine our industries and lifestyles.

Especially after our jaunts through the desert or watching wildlife on hot summer days, every cup of water seems like a blessing.

Settlers

Desert Rats

"I'm not a native, but I got here as fast as I could."
— Seen on a bumper sticker

6.

◀ **Natural pools, such as this one in Sabino Canyon, not only provided settlers with water but also attracted wildlife they could hunt.**
Walt Anderson

▲ **Diegeños, a Yuman-speaking group of Native people living on the far western edge of the Sonoran Desert and over to San Diego.**
from *Report on the United States and Mexican Boundary Survey, Vol. I,* 1857.

▶ **Ancient Hohokam pot, Pueblo Grande Museum, Phoenix.**
Larry Lindahl

The first human settlers, 12,000 to 15,000 years ago, were nomadic hunters and gatherers, who foraged off the land. They may have come across the Bering Strait from Asia. The northern parts of the Sonoran Desert were oak woodland, chaparral, and grassland then.

The archaeological record shows tribes moving in and out of the Sonoran Desert region as the climate warmed and cooled and the resources changed. The mammoths died out about 11,000 years ago, perhaps because of overhunting. People left stone tools, seashells, rock structures, giant ground figures known as intaglios, and foot trails worn into the hard earth. But at times, the climate was so hot and dry that no one lived in big expanses of the region.

Settlements were small, usually family groups camped at water holes or wooded areas. As far as we know, the early tribes had no written language, no wheels, no riding animals, and few possessions. They probably lived in the open, under trees or rock ledges. They built huts with wooden poles, animal hides, and sticks and grass daubed with mud. A few lived in caves, such as Ventana Cave, a rich and revealing archaeological site in Arizona, and some left their artwork in the painted caves of central Baja California.

The Hohokam built a multi-story adobe fortress at Casa Grande, Arizona, and miles of canals along the Salt River at what is now Phoenix. The Trincheras of Sonora made rock walls for protection and for terrace farming of agaves. Native peoples dug wells, erected huts, hunted, farmed, and traded with cultures as far south as what is now Mexico City. They could deliver messages and blue abalone shells from Los Angeles to Phoenix in four days by relay racers covering the 400 miles on foot.

Typical are the Akimel O'odham, who may be descendants of the Hohokam. They settled along the Gila River, raised crops, foraged on native plants, and hunted wild animals. Their neighbors and cousins, the Tohono O'odham, do some farming but prefer ranching.

The first European explorers came to the Sonoran Desert in 1540 looking for cities of gold but found none. Some commented on the bleakness of the area and found the desert harsh and threatening. They met tribes with names such as Cocopah, Maricopa, Yavapai, Quechan, Tohono O'odham, Seri, Dieguito, Cochimi, Kamia, Coahuilla, Yaqui, and Mayo.

Soldiers, priests, miners, farmers, and eventually families followed to settle the land, mine its riches, gather its pearls, graze its grasses. These settlers faced the same problems as the ancient ones: how to grow crops, where to find water, and what to wear for clothing. They established mines, farms, ranches, and villages. They brought riding horses, the wheel, wheat, and cattle. Adobe structures went up, and some of them still stand 300 years later.

Pulses of Humanity

People came in pulses. The California gold rush of 1849 brought waves of prospectors from the eastern United States and Mexico. Others came out of curiosity to study birds and cacti. More came to survey the U.S.–Mexico border. People were drawn to the desert to escape the Depression, to retire, to build airplanes and train for war, or to recuperate from disease in a warm, dry climate. Towns grew into cities and dirt roads became divided highways.

The desert is an austere place that forces reactions from first-time visitors and residents. Olga Wright Smith came west with her husband and father-in-law during the Great Depression to mine for copper in a remote mountain range in southwestern Arizona. They lived in a lean-to shelter propped against boulders. They hauled their water 40 miles from town. The heat reached 110°F much of the summer. At first Olga hated the region, but as she experienced the seasons, bathed in the thunderstorms, made pets of the skunks and bobcats, and saw the flowers in full bloom, she came to love the desert.

When the claim played out and they packed to return to the East, she wrote in *Gold on the Desert:* "I walked in the garden these last days, preferably alone, and looked lovingly at my clumps of brittleweed, my fine beds of desert holly, my borders of paper daisies and desert marigolds... When...I walked the immaculate pathways of the washes, climbed familiar trails up the rocks to look across the wide expanse of sand and sky, or just sat in the sun, never had any place seemed more beautiful than this land I had once feared and despised."

One entrepreneur named Dick Wyck Hall ran a forlorn gas station along a state highway at Salome, Arizona, back when Model-T Fords were new and the nation's highway system was in its infancy. To humor the customers, he invented tall tales of the seven-year-old frog who couldn't swim because it had never seen rain. It had to carry a canteen to keep its back wet. The station dispensed "laughing" gas, and a sign on the wall read, "Drive right up in your old Tin Lizzie, lift up the hood and I'll get busy." Hall's humor kept customers smiling and himself sane in the face of poverty and adversity.

A fellow named Harry Oliver took a liking to the area around Palm Springs, California, between the First and Second World Wars. He sponsored

▲ **Broad-tailed and black-chinned hummingbirds.**

FROM *Report on the United States and Mexican Boundary Survey, Vol. II,* 1859. SELASPHORUS PLATYCERCUS MALE AND FEMALE, TROCHILUS ALEXANDRII MALE.

◀ **Fishhook barrel cactus in bloom.**

LARRY LINDAHL

▶ ***Jesús Varela Marquez and Francisco Cruz Leyva, Rancho Las Palomas, Sonora, 1983***

JAY DUSARD

contests for tall tales about the legendary Pegleg Smith's lost mine, and he even pulled pranks by hiding wooden peg legs in every abandoned mine and prospect he could find. Oliver published the *Desert Rat Scrap Book,* the "only newspaper in America you can open in the wind" (because it was folded from a single sheet of paper). He boasted that it was the world's only five-page newspaper (try it). Oliver explained his journalistic philosophy with, "Your right honourable Editor is indebted to his memory for his jests and his imagination for his facts." People are still laughing at his jokes and wry observations.

Other newcomers to the region were war veterans, retired workers, recluses, or social drifters who found solace and health in the desert. John Butala, an engineer who was gassed in World War I, came to Arizona in hopes his lungs would heal. He squatted on ground south of Ajo, Arizona, built a shack, and went to town once a month for groceries and his pension check. Occasionally, he would do consulting on tough electrical problems for the Phelps Dodge Mining Company. He read technical magazines and let Gambel's quail and cottontails live in his house with him. His long uncut hair and beard led some neighbors to affectionately call him John the Baptist, after the Biblical figure.

Near a cinder mine in the Pinacate region of Mexico, a watchman lived in a one-room tarpaper shack in the mid-1980s. The landscape was black basalt, hot and drab. He loved the place and made it home. Since he had much time on his hands, he made huge piñatas for sale in the distant city. Inside, his shack was filled with a veritable Disneyland of colorful papier-mâché characters: Pinocchio, Mickey Mouse, Dumbo, donkeys, roses. His creations made thousands of children happy.

Such unexpected and unique characters led to the term "desert rat" for people who love living here. It is a term of endearment, just as "sidewinder" or "vulture" are insults. Edward Abbey, author of many books on the Southwest, was a desert rat of the first rank, but when camping in a place as rough as the

▲ **Norwegian explorer Carl Lumholz extolled the desert as "radiant with good cheer."**

CARR CLIFTON

Pinacate lava he wondered if he were a true desert rat or just another desert mouse.

Norwegian explorer Carl Lumholtz had seen the world but became addicted to the Sonoran Desert. He spoke for a host of people when he wrote in *New Trails in Mexico,* "To me the desert is radiant with good cheer; superb air there certainly is, and generous sunshine, and the hardy, healthy-looking plants and trees with their abundant flowers inspire courage. One feels in communion with nature and the great silence is beneficial."

Any time millions of people move into a region, the land changes. It's a story as old as civilization. In our quest for water we dam rivers to irrigate crops, control floods, and have something to drink. Those same dams crank electricity to distant cities. Corridors of cottonwoods and mesquite forests are cut to make firewood, to gain lumber for houses and fences, or to open land for farms. Photos from the 1850s show immense woodpiles stacked along the lower Colorado River to fuel the boilers of steamboats, but today remnants of the forests that supplied that wood are seen in only a few places.

In plains and valleys away from the rivers, we drilled wells and built windmills to pump water. Water tables dropped. Mesquite bosque forests died, entire species of native fish died, herds of pronghorn were evicted, trees and creosote flats were cleared—today bulldozers gobble acres an hour. Running out of water remains the biggest fear and challenge for desert communities.

Newcomers battle the enchantment of a new place with a nostalgic desire to remake it like "back home." It takes time to discover a sense of place, and on occasion we are lost in a sea of look-alike houses and franchise businesses, so that one desert town looks like all towns from coast to coast. We wonder why we live here instead of somewhere else. Why do we stay? Is this home?

Eventually, like Olga Wright Smith, we come to expect the arrival of white-winged doves, we long for the song of the coyote, we lust to see a saguaro blooming. We learn to love the desert, call it home, and defend it.

It is far easier to protect a place and care for it than to fix it after it's ruined. In many places of the world, human overuse has created dead zones where the trees and animals are gone. Land is cleared, wells dry up, farmlands harden with salts, crops die, and soil blows away on the next wind. Land once vibrant and alive turns barren and worthless.

Ultimately, we understand and support the need to protect what we value. We can be kinder to our land. We can create protected places of public land—parks, monuments, wilderness areas, refuges—so that we all can enjoy nature. For example, four new national monuments recently were dedicated in the Sonoran Desert, and a seven-million-acre peace park along Arizona's border with Mexico may be next.

On our private land we don't need to chop down every tree and burn every cactus when we build a house or factory. We don't need to eradicate every insect or spider, every lizard and pocket mouse. We learn that saving the desert is more cost-efficient, and certainly more satisfying. Saving some desert for the critters is good for them and us.

Native wildlife and plants add much to our lifestyle. Take the cactus wren. It starts its day by searching every wall and windowsill for insects. It makes nests in cholla cacti. It has no bad habits. The cholla cacti and mesquite trees it lives in are already adapted to the sparse rainfall and hot summers. And its calls make us smile. What's not to like?

▶ **Pyrrhuloxia.**
Tom Vezo

▼ **Though tarantulas may look fearsome, they are actually quite benign and beneficial.**
Tom Vezo

With a little consideration, we can see much the same wildlife at home that we pay to see in parks and zoos. It's mostly a matter of keeping our own cats and dogs in check and of being a little tolerant if the howls of the coyote wake us up, the bull snake startles us on the patio, the javelinas root up a few plants, or the Gila woodpecker bangs on the wall too early in the morning.

Urban wildlife requires the same things that we do: shelter, food, water in some form, and room to roam looking for a mate. By leaving some ground bare we provide places for quail to dust-bathe and for pocket mice and antelope squirrels to burrow. By leaving creosote and bursage bushes in place, we encourage jackrabbits and cottontails. By planting native milkweeds and passion-flower vines, we attract butterflies.

Many people put out seed for quail, doves, and cardinals. Sugar-water feeders for hummingbirds bring them to our kitchen windows year-round. Small ponds, birdbaths, and drips provide viewing stations. Backyard wildlife is fun—like a self-guiding, low-cost private zoo.

A covey of quail came by a few minutes ago, clucking and scratching for seeds. The creosote bush is blooming again for the third time this year, and the ocotillo by the door is tipped with fiery red flowers. A tree lizard hangs on the wall waiting for a beetle to land. Nature's gifts are priceless yet free.

Few things are more memorable than hiking under a full moon, sleeping under the stars, walking hand-in-hand on the beach, watching a bighorn lamb scamper up a cliff, inhaling the fragrance of a night-blooming cereus, or discovering a secret canyon. Foremost, the Sonoran Desert is alive. And lest we sometimes forget, so are we.

Pull up a chair in the shade and we'll check what's in the ice chest. Tomorrow we can see what's over the next hill.

◀ **Ocotillos at dawn, Cabeza Prieta National Wildlife Refuge.**

Jack Dykinga

SOURCES

Chapter 1

Edmonson, Travis J. "The Purpose of the Desert," *Thoughts That Didn't Pass* (Tucson: private printing, 2001), p. 11.

Courtenay, Bryce. *The Power of One* (New York: Ballantine, 1996), pp. 154–155.

McKasson, Molly. "Molly's Desert Journal," *Tucson Guide Quarterly,* vol. 14, no. 3 (fall 1996), p. 25.

Chapter 3

Cummings, E. E. "maggie and millie and molly and may," No. 10, *95 Poems* (New York: Harcourt Brace Jankovich, 1972), p. 682.

Lumholtz, Carl. *New Trails in Mexico* (London: T. Fisher Unwin, 1912). Reprinted 1971, Glorieta, New Mexico: Rio Grande Press, pp. 223–224.

Saint-Exupéry, Antoine de. *Wind, Sand and Stars.* (New York: Harcourt, 1992), p. 85.

Chapter 4

Manje, Juan Mateo. *Luz de Tierra Incógnita* (Tucson: Arizona Silhouettes, 1954), p. 161.

Wright, Franz. "Year One." *The New Yorker* (November 26, 2001), p. 59.

Chapter 5

Mearns, Edgar Alexander. *Mammals of the Mexican Boundary of the United States* (Washington, D.C.: Smithsonian Institution, 1907). Reprinted, New York: Arno Press, 1974, pp. 128–129.

Chapter 6

Smith, Olga Wright. *Gold on the Desert.* (Albuquerque: University of New Mexico Press, 1956), pp. 246–247.

Lumholtz, Carl. *New Trails in Mexico* (London: T. Fisher Unwin, 1912). Reprinted 1971, Glorieta, New Mexico: Rio Grande Press, p. 307.

FOR FURTHER ENJOYMENT

Burckhalter, David. 1999. *Among Turtle Hunters & Basket Makers: Adventures with the Seri Indians.* Tucson: Rio Nuevo.

Heisey, Adriel. 2000. *Under the Sun: A Sonoran Desert Odyssey.* Tucson: Rio Nuevo.

Humphreys, Anna, and Susan Lowell. 2002. *Saguaro: The Desert Giant.* Tucson: Rio Nuevo.

SEEING THE SONORAN DESERT

Seeing the Sonoran Desert is easy. We're seldom more than an hour's drive from a park or museum devoted to it. Addresses and phone numbers for these prominent places can be found using the Internet, phone book, or government directories.

Key to abbreviations

BLM	Bureau of Land Management
FS	Forest Service
FWS	Fish & Wildlife Service
NPS	National Park Service
NWR	National Wildlife Refuge

Museums, Zoos & Gardens

Arizona-Sonora Desert Museum.
520-883-1380, Tucson, Arizona

Boyce Thompson Arboretum.
520-689-2811, Superior, Arizona

CEDO (Centro Intercultural de Estudios de Desiertos y Océanos). Puerto Peñasco, Sonora, Mexico

Centro Ecológico.
Hermosillo, Sonora , Mexico

Desert Botanical Garden.
480-941-1225, Phoenix, Arizona

The Living Desert Zoos & Gardens.
760-346-5694, Palm Desert, California

The Phoenix Zoo.
602-273-1341, Phoenix, Arizona

Pioneer Arizona Living History Museum.
623-465-1052, Phoenix, Arizona

Tohono Chul Park.
520-544-7922, Tucson, Arizona

Tucson Botanical Gardens.
520-326-9255, Tucson, Arizona

National Parks & Monuments (U.S.)

Agua Fría National Monument. BLM, 623-580-5500, Phoenix, Arizona

Casa Grande Ruins National Monument. NPS, 520-723-3172, Casa Grande, Arizona

Ironwood Forest National Monument. BLM, 520-258-7200, Tucson, Arizona

Joshua Tree National Park. NPS, 760-367-5500, Twentynine Palms, California

Organ Pipe Cactus National Monument. NPS, 520-387-6849, Tucson, Arizona

Saguaro National Park. NPS, 520-733-5158 (west unit), 520-733-5153 (east unit), Tucson, Arizona

Santa Rosa and San Jacinto National Monuments. BLM, FS, and others. Palm Springs, California

Sonoran Desert National Monument. BLM, Phoenix, Arizona

National Wildlife Refuges (U.S.)

Bill Williams River National Wildlife Refuge. FWS, 928-667-4144, Parker, Arizona

Cabeza Prieta National Wildlife Refuge and Wilderness. FWS, 520-387-6483, Ajo, Arizona

Havasu National Wildlife Refuge. FWS, 760-326-3853, Needles, California

Imperial National Wildlife Refuge. FWS, 928-783-3371, Yuma, Arizona

Kofa National Wildlife Refuge and Wilderness. FWS, 928-783-7861, Yuma, Arizona

Sonny Bono Salton Sea National Wildlife Refuge. FWS, 760-348-5278, Calipatria, California

State Parks (U.S.)

Alamo Lake State Park. 928-669-2088, Wenden, Arizona

Anza-Borrego Desert State Park. 760-767-5311, Borrego Springs, California

Anza-Borrego Sky Trail. 760-767-5311, Borrego Springs Airport, California

Buckskin Mountain State Park. 928-667-3231, Parker, Arizona

Catalina State Park. 520-628-5798, Tucson, Arizona

Cattail Cove State Park. 928-855-1223, Lake Havasu City, Arizona

Heber Dunes State Vehicle Recreation Area. 760-767-5391, El Centro, California

Indio Hills Palms State Park. 760-393-3059, Palm Springs, California

Lake Havasu State Park. 928-855-2784, Lake Havasu City, Arizona

Lost Dutchman State Park. 480-982-4485, Apache Junction, Arizona

Ocotillo Wells State Vehicle Recreation Area. 760-767-5391, Borrego Springs, California

Picacho Peak State Park. 520-466-3183, Picacho, Arizona

Picacho State Recreation Area. 760-393-3052, Winterhaven, California

Salton Sea State Recreation Area. 760-393-3052, North Shore, California

Yuma Crossing State Historic Park. 928-329-0471, Yuma, Arizona

County & City Parks (U.S.)

Besh-Ba-Gowah Archaeological Park. 520-425-0320, Globe, Arizona

McDowell Mountains Regional Park. 602-471-0173, Fountain Hills, Arizona

Pueblo Grande Museum and Cultural Park. 602-495-0901, Phoenix, Arizona

South Mountain Park. 602-534-6324, Phoenix, Arizona

Spur Cross Ranch Conservation Area. 480-488-6601, Scottsdale, Arizona

Tucson Mountain Park. 520-740-2690, Tucson, Arizona

White Tank Mountain Regional Park. 623-935-6056, Phoenix, Arizona

Wilderness & Special Use Areas (U.S.)

Aravaipa Canyon. BLM, 928-348-4400, Mammoth, Arizona

Eagletail Mountains Wilderness Area. BLM, 928-317-3200, Phoenix, Arizona

Imperial Sand Dunes Recreation Area. BLM, El Centro, California

New Water Mountains Wilderness Area. BLM, 928-317-3200, Yuma, Arizona

North Algodones Dunes Wilderness. BLM, 760-337-4400, El Centro, California

Sabino Canyon Recreation Area. FS, 520-749-2861, Tucson, Arizona

Federal Biosphere Reserves, Protected Areas & Parks (Mexico)

Area de Protección de Flora y Fauna, Islas de Golfo de California. La Paz, Baja California Sur

Area de Protección de Flora y Fauna, Valle de los Cirios. Guerrero Negro, Baja California Sur

Area Protegida Cajón del Diablo. Guaymas, Sonora

Parque Nacional Bahía de Loreto. Loreto, Baja California Sur

Parque Nacional Cabo Pulmo. San José del Cabo, Baja California Sur

Parque Nacional Constitución de 1857. Baja California

Parque Nacional Sierra de San Pedro Mártir. San Felipe, Baja California

Reserva de la Biosfera Alto Golfo de California y Delta del Río Colorado. San Luis Río Colorado, Sonora.

Reserva de la Biosfera El Pinacate y Gran Desierto de Altar. Puerto Peñasco, Sonora.

Reserva de la Biosfera El Vizcaíno. Guerrero Negro, Baja California Sur

Reserva de la Biosfera Ojo de Liebre. Guerrero Negro, Baja California Sur

▲ **Ocotillo in bloom, East Cactus Plains Wilderness, Arizona.** David H. Smith

ACKNOWLEDGMENTS

I'd need more space than this book has to thank everyone who has tried to teach me about the desert or who has encouraged me to put thirsty words on paper. Some of the many who generously have shared camps or conversations include: Ed Abbey, John Annerino, Byrd Baylor, Jayne Belnap, Chuck Bowden, Diane Boyer, Charles Conner, Roger DiRosa, Dan Duncan, Chuck Fellows, Jack Dykinga, Luke Evans, Don Fedock, Richard Felger, Bunny Fontana, "Cactus Kate" Garmise, "Petey Mesquitey" Gierlach, Byrd Granger, Trish and David Griffin, John Gunn, Jim Gutmann, Gayle and Bill Hartmann, Julian Hayden, Freeman Hover, Bill Hoy, Ronald Ives, Gene Joseph and Jane Evans, Tom Kleespie, Chuck and Marilyn Kline, Sandy Lanham, Peggy and Merv Larson, Joe McCraw, Al McGinnis, Nonie McKibbin, Gale Monson, Don Neff, Ami Pate, Doug Peacock, Dave Roberson, Sue Rutman, Bob Schumacher, Cecil Schwalbe, Fran Sherlock, Norman Simmons, Betty and Dana Smith, Greg Starr, Dale Turner and Julia Fonseca, Ray and Jeanne Turner, Sharon Urban, Dan Urquidez, Phil Varney, Joe Wilder, Jeanne Williams, Bill and Beth Woodin, David Yetman, and Ann Zwinger. These are my heroes.

Thanks to the legion of researchers and explorers who follow their curiosities into the unknown wilds of biology, geology, archaeology, history, geography, and ecology, and then tell us what they found.

Special thanks to Lisa Cooper, Ron Foreman, Janice Harayda, Susan Lowell Humphreys, and Ross Humphreys at Rio Nuevo Publishers.

ABOUT THE AUTHOR

BILL BROYLES, a research associate at the University of Arizona Southwest Center, is a student of desert places. He served in the high school classroom for 31 years, teaching English and physical education. Currently he is working to create a Sonoran Desert peace park on the Arizona-Sonora border. His written work has appeared on the pages of such publications as *Arizona Highways, Journal of the Southwest, Journal of Arizona History, Journal of Arid Environments, Wildlife Society Bulletin,* and the book *Organ Pipe Cactus National Monument: Where Edges Meet.* He prefers long, solo hikes across the valleys, dunes, and mountains, but he also loves sitting under a tree to read a book, watch desert bighorn, or search the horizon.

RIO NUEVO PUBLISHERS
P.O. Box 5250, Tucson, AZ 85703-0250
(520) 623-9558
www.rionuevo.com

Editors Janice Harayda and Ronald J. Foreman

Design Larry Lindahl

Cover Design David Jenney

Library of Congress Cataloging-in-Publication Data
Broyles, Bill.
Our Sonoran Desert / Bill Broyles.
p. cm.
ISBN 1-887896-40-6 (pbk. : alk. paper)
1. Natural history—Sonoran Desert. 2. Sonoran Desert. I. Title.
QH104.5.S58B77 2003
508.3154'09791'7—dc21

2002152299

Printed in Korea

10 9 8 7 6 5 4 3

FRONT COVER: Teddy bear cholla and brittlebush, Organ Pipe Cactus National Monument. LARRY ULRICH

TITLE PAGE: Sunrise over badlands, Anza-Borrego Desert State Park. SCOTT T. SMITH

BACK COVER: Saguaro and brittlebush, Organ Pipe Cactus National Monument. LARRY ULRICH

CONTENTS: Cactus wren (right). TOM VEZO
Vizcaíno Reserve (upper left). GEORGE H. H. HUEY
Petroglyphs, Saguaro National Park (lower left). LARRY LINDAHL

In this land of contrasts, desert mountains give way to a mangrove swamp in an estuary north of Punta Chueca, Sonora, Mexico, in Seri country. MILLS TANDY